An Introduction to Cell Biology

An Introduction to Cell Biology

C B Hole

Macmillan

First published 1974

Published by
MACMILLAN EDUCATION LTD
London and Basingstoke
Associated companies and representatives
throughout the world

Printed and bound in Great Britain at The Pitman Press, Bath

Preface

This guide book is designed to lead the student into an exciting and expanding area of Biology. It is not an 'A' level textbook but is intended for use in pre-lesson preparation, post-lesson consolidation and in reducing the amount of classroom time spent in making notes. It merely sketches the bare bones of cell biology by providing basic information from the fields of cytology, organic chemistry and biochemistry. Each teacher or student will want to amend, enlarge and embellish these bones during classroom teaching, discussion, practical lessons and further reading to suit his own needs and interests.

Since chemistry is often a stumbling block for biologists, biochemical processes are treated very simply so that students with a poor chemical background can understand them. It has not been possible to illustrate this book with photographs or coloured diagrams nor has it been possible to discuss the important research which led, and still leads, to great advances in this field. Photographs, however, are already available inexpensively and experimental evidence is extensively documented elsewhere so it is felt that these omissions are justified in the interests of economy.

Contents

Diagrams and Tables
Common Units

1   **Introduction:** Science and Cells   11

2   **Cell Structure:** A Generalised Cell   13

3   **Cell Chemistry**   28
Organic macromolecules. Carbohydrates.
Proteins. Lipids.
Nucleic acids. Enzymes.

4   **Energy Conversion in the Cell**   60
Energy Conversion. Photosynthesis. Respiration.

5   **The Cell at Work 1**   78
Controlling the internal environment. Biosynthesis: protein synthesis,
nucleic acid synthesis, carbohydrate synthesis, lipid synthesis.

6   **The Cell at Work 2:** Growth and Division   90

7   **Cell Types**   94
Procaryotes and eucaryotes. Plant and animal cells.
Specialised cells: the muscle cell, the nerve cell, the gametocyte.

8 Viruses — The Cell Parasites   105
Virus classification. Viruses in action. The temperate phage. Viral genetics.

9 The Cell in Review   112
The cell as a whole. The tasks ahead.

   Index   119

Diagrams and Tables

2 *Diagrams*
2.1 A scale of size in cell biology   14
2.2 A generalised eucaryotic cell   15
2.3 The cell wall   16
2.4 The plasma membrane   17
2.5 A mitochondrion   19
2.6 A chloroplast   22
2.7 A centriole   24
Tables
2.1 Cell components   20

3 *Diagrams*
3.1 Monosaccharides: triose sugars, pentose sugars   31
3.2 Monosaccharides: hexose sugars   34
3.3 Disaccharides   36
3.4 Polysaccharides: chains of rings   38
3.5 Phosphorylated monosaccharides   39
3.6 Proteins: amino-acids   41
3.7 Types of protein bond   43
3.8 Protein structure   44
3.9 Lipids: fatty acids, fats and oils   46
3.10 Lipids: phospholipids, steroids   47
3.11 Nucleic acid components: purines and pyrimidines   50
3.12 Nucleic acids: nucleotide chain, DNA chain   53
3.13 Nucleic acids: the base pairs   54
3.14 Enzymes: action, inhibition   57
3.15 Enzyme activity factors   58
Tables
3.1 Organic macromolecules — construction patterns   29
3.2 Monosaccharides: polymerised carbohydrates   33

Diagrams and Tables

4 *Diagrams*
4.1 Autotrophs and heterotrophs   61
4.2 The cell in action: energy conversion   62
4.3 Photosynthesis: basic input/output reaction, backbone reactions   63
4.4 Photosynthesis — detailed reactions.   1 Photolysis   65
4.5 Photosynthesis — detailed reactions   2 Carbon dioxide fixation   67
4.6 Photosynthesis — detailed reactions   3 Pathway of carbohydrate conversion   68
4.7 Respiration: basic input/output equations, backbone diagram   70
4.8 Respiration details   1 Glycolysis   71
4.9 Respiration details   2 Krebs cycle   73
4.10 Respiration details   3 Respiratory chain of hydrogen carriers   74
4.11 Respiration details   4 Fuels for respiration   76

5 *Diagrams*
5.1 The carrier concept in active transport   81
5.2 Role of DNA in biosynthesis   82
5.3 Coding patterns: DNA to protein   85
5.4 Coding patterns: DNA to DNA   87
5.5 Gene mutations produced by insertion or deletion of nucleotide   88
Tables
5.1 Types of nucleic acid found in cell   83
5.2 Plan of protein synthesis   84

6 *Diagram*
6.1 The cell cycle   91

7 *Diagrams*
7.1 The muscle cell   97
7.2 The nerve cell   99
7.3 The nucleus in division: mitosis and meiosis   101
Tables
7.1 Characteristics of procaryotic cells and eucaryotic cells   95
7.2 Characteristics of plant cells and animal cells   96

8 *Diagrams*
8.1 Virus types   107
8.2 Viruses in action: a bacteriophage attack   108
Tables
8.1 Viruses: relative size scale   105
8.2 Classification system for viruses   106

9 *Diagrams*
9.1 Patterns of cell activity   113
9.2 Jacob and Monod 'operon' theory   116

Common Units

SI units

1 m	metre (m)	= 1000 mm	1m	⎫ human
0·1 m				⎬ eye
0·01 m				⎭ range
0·001 m	millimetre (mm) = 1000 μm		10^{-3} m	⎫
0·0001 m				light
0·00001 m				microscope
0·000001 m	micrometre (μm) = 1000 nm		10^{-6} m	range
0·0000001 m				⎫ electron
0·00000001 m				⎬ microscope
0·000000001 m	nanometre (nm)		10^{-9} m	⎭ range

Relation to traditional units

1 micron (μ)	=	1 micrometre (μm)
1 millimicron (mμ)	=	1 nanometre (nm)
1 Ångstrom unit (Å)	=	10^{-1} nanometre or 10^{-10} metre

The traditional units on the left are found in older British books and in American publications.

1 Introduction: Science and Cells

Scientific progress involves an interplay of ideas and investigation. Observations lead to ideas, ideas lead to theories, theories lead on to further investigation for which better tools are developed. New observations become possible with improved instruments and so fresh ideas are generated. The development of cell biology provides a good illustration of this concept.

The ancient Greeks, insatiably curious, observed and theorised about the living world but their view was limited by their lack of magnifying power. Scientific curiosity, unpopular through the Dark Ages, flourished again in the seventeenth century. New attitudes encouraged enquiry and scientists entered the laboratory where they developed special instruments to use in their investigations.

The ingenious Dutch draper, Anton van Leeuwenhoeck, made simple microscopes at this time using glass lenses ground in his own workshop. A long series of letters written by him to the Royal Society of London describes in amazing detail the micro-organisms or 'animalcules' he found living in pond water, teeth scrapings and other fertile fluids. A contemporary, Robert Hooke, an Englishman and a versatile scientist, designed a compound microscope and recorded his microscopic observations in a book called *Micrographia*, published in 1664. Here the word 'cells' was first used when Hooke described small pores in a slice of cork.

Other scientists using microscopes to examine biological material also reported pores like Hooke's 'cells'. These observations slowly accumulated and eventually led to a theory postulating that all organisms are composed of cells. This 'Cell Theory', published in 1839, was based upon the work of two Germans, the botanist Schleiden and the zoologist Schwann.

In the early years of the nineteenth century the development of achromatic lenses reduced colour distortion in the compound microscope, allowing greater magnification to be produced. Observations with these improved microscopes supported the Cell Theory and allowed Rudolph Virchow to infer 'omnis cellulae cellula' — that new cells are derived only from pre-existing cells — an inference which, though a commonplace today, startled Virchow's contemporaries.

The contents of the cell began to receive attention about this time. The cell nucleus was described by Robert Brown in 1831. By 1879 Walter Flemming, using newly-developed aniline dyes which differentially stained nuclear material, was able to follow the sequence of nuclear events at cell division.

Further technical advances have been made in the twentieth century. Phase-contrast and interference microscopes, refinements of the optical microscope, increase the contrast between objects with different refractive indices and so reveal more detail in thin unstained specimens. These techniques are particularly valuable for use with living material. And then, in the 1930s, a completely new type of microscope was developed. This 'electron microscope' bypasses the physical limitations imposed upon optical magnification by the wavelength of light. Instead of the conventional light source it uses an electron beam of shorter wavelength and electromagnetic 'lenses' replace the traditional optical lens system. Images of specially preserved and treated specimens, magnified up to x 200 000, can be produced on a fluorescent screen or a photographic plate and objects as small as 1 nm can be resolved.

Since the middle of this century these new techniques, combined with others developed in the allied field of biochemistry, have led to great advances in biology. Investigation has centred on the structure and functions of Hooke's cell — the basic unit of living things — and the results of twenty years' productive research have provided the material for this book.

2 Cell Structure: a Generalised Cell

Introduction

Cells come in a wide variety of shapes, sizes and component parts. Environmental and functional differences dictate structural diversity. Many cells exist as free-wheeling, unicellular organisms which must be wholly self-sufficient, while others are bound in the conglomerates of plant and animal tissues where they may become highly specialised and heavily dependent upon the functions of other cell types.

This range of structural variety is now brilliantly displayed by the work of electron microscopists. Micrographs cannot be included in such a small notebook but they should not be missed; diagrams compounded from them, though useful, give little indication of the fascinating detail visible in the originals (see references at the end of this chapter).

The generalised cell

The wide diversity of cell types conceals an essential conformity, so that similarities at the biochemical level of cellular structure and function are more striking than differences. It is valid, therefore, to discuss the structure of a mythical 'generalised cell' and then to consider how actual cell types deviate from this hypothetical composite.

The structure of a generalised cell

A cell is usually made up of a central nucleus embedded in cytoplasm and wrapped with a flexible plasma membrane coat which may be further enclosed by a rigid cell wall. Most or many of the following cellular components will be present in any particular cell type.

1 Cell wall (found mainly in fungi, bacteria and plant cells)
2 Plasma membrane
3 Cytoplasm, containing:

(a) endoplasmic reticulum
(b) Golgi apparatus
(c) mitochondria
(d) chloroplasts
(e) ribosomes

(f) lysosomes
(g) centrioles
(h) vacuoles
(i) granules
(j) microtubules and microfilaments

2.1   *A scale of size in cell biology*

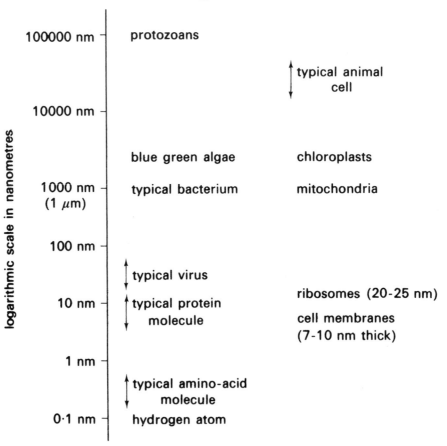

4 Nucleus, made up of:
 (k) nuclear membrane
 (l) nucleoli
 (m) chromosomal material

2.2   *A generalised eucaryotic cell (based upon electron micrographs of thin sections)*

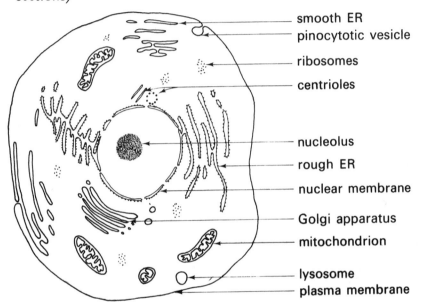

smooth ER
pinocytotic vesicle
ribosomes
centrioles

nucleolus
rough ER
nuclear membrane
Golgi apparatus
mitochondrion
lysosome
plasma membrane

1 The cell wall

Every cell is delimited from its environment by a plasma membrane but many cells also possess a cell wall which lies outside the membrane.

A complex outer cell wall made of the carbohydrates cellulose and pectin is a characteristic of most plant cells. It forms a permeable framework which strengthens and protects plant tissues. The many layers of the cell wall develop as the plant cell grows. First the extensible, cellulose primary wall is laid down between the plasma membrane and the middle lamella (an inter-cellular cement of pectin found in plant tissues). Then, as cell growth ceases, the layered secondary wall is deposited on the inner surface of the primary wall (see diagram 2.3). The cellulose fibres of each successive layer lie at different angles, increasing the strength of the cell wall in a manner analogous to the reinforcing effect of fibre layers placed at different angles in the wall of a car tyre. Lignin may be added to make a rigid secondary wall when the greater strength of woody tissue is required.

2.3   *The cell wall. A mature plant cell showing the development of 1) a primary cell wall and 2) a secondary cell wall.*

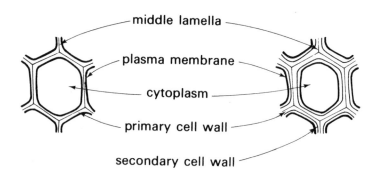

middle lamella

plasma membrane

cytoplasm

primary cell wall

secondary cell wall

1   2

Animal cells do not have true cell walls but some are enveloped, wholly or partially, by wall-like protective pellicles such as the chitin pellicles of insects and arthropods and the keratin layers of some vertebrate cells.

The cells of tissues are often bound together by an intercellular matrix or layer which may serve several purposes. Plant cells are cemented together by the middle lamellae, and the matrix of bone and cartilage tissues contributes largely to the supporting function of these tissues as a whole.

2 The plasma membrane (cell membrane)

All cells are bounded by a lipoprotein plasma membrane about 10 nm thick. This flexible membrane surrounds and supports the cell, controlling the entry and exit of materials by means of its selective permeability (see page 78).

The lipid constituents of the membrane (phospholipids, glycolipids and the sterol cholesterol) play a structural role and their relative amounts are very variable. The phospholipid molecules are usually arranged in two layers so that the non-polar, water-insoluble fatty acid tails (see page 46) lie within the membrane and the polar, water-soluble molecular heads make up the internal and external surface layers (see diagram 2.4). The protein molecules act largely as enzymes and are responsible for specific functions of the membrane such as molecular transport. Many different types of protein

2.4   *The plasma membrane*

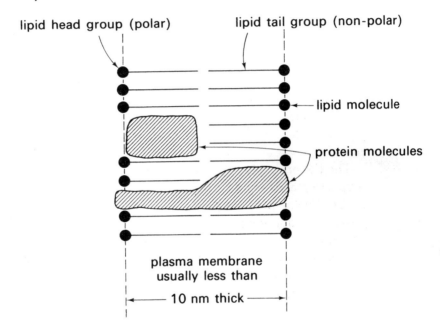

are arranged irregularly on or within the membrane surface. This two-layered lipid structure peppered with varying amounts of protein is often called a 'unit' membrane and it is characteristic of membranes found in the cell.

The plasma membrane may form a simple cellular envelope or it may be adapted in several ways. Absorptive cells develop finger-like projections of the plasma membrane, microvilli, which greatly increase their surface area. The protective, insulating myelin sheath of nerve fibres is formed from the plasma membranes of Schwann cells which extend to wrap spirally round the nerve axon. The plasma membranes of neighbouring cells in a tissue may develop special areas of attachment such as the desmosome which involves thickening of the adjacent membranes from which intracellular filaments radiate.

3 Cytoplasmic constituents

The cytoplasm is packed with an orderly array of membranes and organelles.

(a) The endoplasmic reticulum

The endoplasmic reticulum (ER) is an elaborate, sac-like system of lipo-protein 'unit' membranes lying within the cytoplasm. This complex extends in varying degrees from the plasma membrane to the nuclear membrane. There are two types of ER:

   (i) rough or granular ER, which is heavily coated with ribosomes (see page 23) on its cytoplasmic surface,

   (ii) smooth ER, which is not associated with ribosomes.

   Although most cells contain both rough and smooth ER their quantity and relative proportions may vary widely. Rough ER occurs extensively in protein-secreting cells such as those of the mammalian salivary gland and pancreas whereas well-developed smooth ER is found in steroid-producing cells.

   The ER probably has many important functions. Primarily it serves as a supporting platform for the ribosomes of the rough ER and as a transport system for the proteins produced by them but the membranes of the ER also provide conducting pathways for the movement of many other materials within the cell. The ER forms a structural framework which divides the internal volume of the cell into cytoplasmic compartments and provides surfaces upon which chemical reactions may take place in regular sequence without mutual interference.

(b) The Golgi apparatus (Golgi complex or Golgi body)

The Golgi apparatus, another elaboration of the cytoplasmic membranes, is a complex of flattened membranous sacs with a cluster of vesicles at their edges. This apparatus has been found almost universally in eucaryotic cells (all cells except those of bacteria and blue-green algae, see page 94) though its size and position is variable.

   The Golgi apparatus is involved in the secretion and packaging of complex carbohydrates and proteins; it is very conspicuous in actively secreting cells. Polysaccharides are thought to be both synthesised and secreted here whereas proteins are only packaged for export in the apparatus, being synthesised by the ribosomes of the rough ER (see page 23).

(c) *Mitochondria*

Mitochondria are membrane-bounded organelles found in all eucaryotic cells.

2.5   *A mitochondrion*

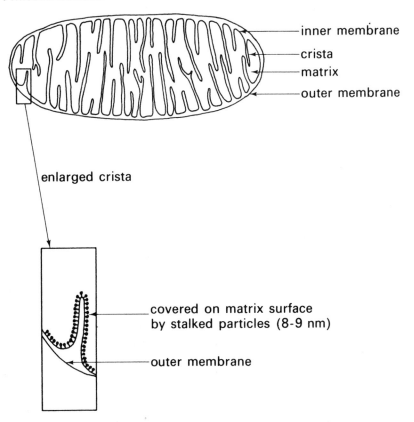

They are cellular 'power houses' where aerobic respiration provides a supply of utilisable energy in the form of ATP molecules (adenosine triphosphate, see page 60). While most cells contain at least 200 mitochondria, their number and arrangement depend upon the cell's energy needs. Cells with high energy requirements contain very large numbers of mitochondria concentrated at the site of energy utilisation.

Mitochondria can be spherical, sausage-shaped or branched and they vary widely in size (roughly 500 nm diameter × 2000 nm long). The inner of the two surrounding membranes is thrown into folds, or cristae (see diagram 2.5), which project towards the matrix and provide a large internal surface area. The lining of the matrix may be covered with small stalked granules.

Each mitochondrion contains DNA (see page 47) and ribosomes and can synthesise some of its own proteins. These organelles are self-replicating, arising only by division from pre-existing mitochondria, and it is thought that their DNA contains much of the hereditary genetic information needed to control mitochondrial growth and reproduction although the cell must supply some enzymes and other essential molecules. This unusual degree of autonomy has led to the suggestion that mitochondria are descended from free-living, procaryotic organisms (similar to bacteria, see page 94) which became incorporated into larger cells as endosymbionts during evolution.

Cellular respiration takes place largely within the mitochondria and this process will be discussed in detail in a later chapter (see page 68). Different parts of the process take place in specifically designated areas of the mitochondria.

Table 2.1   *Cell components*

Structure	Shape and size	Location	Function
Cell wall	Variable	Outside plasma membrane, mainly in plant cells	Support, protection and special functions
Plasma membrane	Lipoprotein 'unit' membrane 8 – 10 nm thick	Outer surface of all cells	Limiting membrane, controls traffic of materials into and out of the cell
Endoplasmic reticulum (ER)	Lipoprotein membrane (a) rough ER with ribosomes (b) smooth ER no ribosomes	Throughout cytoplasm	Membrane contains enzymes and provides reaction surfaces. Acts as conducting and compartmenting system. Rough ER associated with protein synthesis, smooth ER involved in steroid synthesis
Golgi apparatus	Stack of membranous sacs, variable size and shape	In cytoplasm	Manufacture of complex polysaccharides, concentration and packaging of these and proteins

Structure	Shape and size	Location	Function
Mitochondria	Sausage-shaped, approximately 500 × 2000 nm	Few to 1000 in cytoplasm, more in active cells	Powerhouse of cell, breaking down carbohydrates to release energy
Chloroplasts	Egg or disc-shaped, about 5000 – 10000 nm in diameter	In cytoplasm of green plant cells	Site of photosynthesis, light energy trapped by chlorophyll to make carbohydrates from CO_2 and H_2O
Ribosomes	Spherical, about 20 – 25 nm diameter	Free in cytoplasm or associated with rough ER	Site of protein synthesis
Lysosomes	Spherical, about 500 nm diameter	In cytoplasm of many cells	Cell disposal units containing digestive enzymes
Centrioles	Complex, rod-like structures 200 nm diam × 400 nm long	Members of centriole pair lie at right angles to each other near nuclear membrane in cytoplasm of animal cells	Each member of the pair moves to form a pole of the nuclear spindle during cell division
Vacuoles	Variable	Prominent in cytoplasm of plant cells	Maintain turgidity and contain food or waste materials
Granules	Variable	In cytoplasm	Variable, mainly storage and excretory
Micro-filaments and tubules	Rods and tubes 5 – 20 nm diameter	In cytoplasm	Variable
Nuclear membrane	Two 'unit' membranes with small pores	Surrounds nucleus	Governs traffic of materials into and out of the nucleus
Nucleolus	Spherical and dense, not enclosed by membrane	One to four in each nucleus	Site of synthesis of ribosomal RNA
Chromosomal material (chromatin)	Dispersed as fine DNA/protein strands in active cells, coiling to form chromosomes during division	In nucleus	Chromosomes govern cell activities by their hereditary instructions for protein synthesis which are coded into DNA molecule.

(d) Chloroplasts

Chloroplasts are intricate organelles found in some plant cells. Light energy can be captured by the green chlorophyll pigment of chloroplasts and used to power the process of photosynthesis whereby complex food molecules are built from inorganic raw materials (see page 63).

2.6   A chloroplast

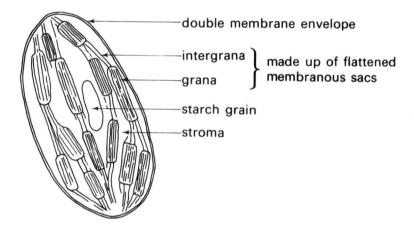

Three grana

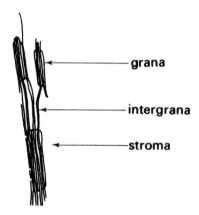

A chloroplast is a large structure (approximately 5000—10 000 nm in diameter) usually shaped like a convex lens and surrounded by two unit membranes. Internally it is made up of discrete areas, the grana, contained in a ground substance, the stroma. The grana are composed of disc-like membranous sacs called lamellae stacked up in piles. Lamellae extend through the stroma to provide continuity between neighbouring grana (see diagram 2.6). Layers of protein form the lamellar membrane surfaces and sandwich between them layers of chlorophyll, enzymes and phospholipids. The chloroplasts of algae show variations from this structure which is characteristic of higher plants.

   Chloroplasts, like mitochondria, possess their own supply of DNA and ribosomes. They are derived from pre-existing chloroplasts by division, developing from undifferentiated proplastids by growth and elaboration of the lamellar pigments and membranes. Chlorophyll does not develop if chloroplasts are grown in the dark but it rapidly appears on exposure to light.

(e) Ribosomes

Ribosomes are very small, dense particles (diameter 20–25 nm) found in the cytoplasm where they may lie free or be closely associated with the ER. They are universally present in cells.

   Ribosomes are the site of protein synthesis. They are rich in RNA which is manufactured in the nucleoli (see page 26). Free ribosomes are thought to produce the proteins required for internal cellular use whereas ribosomes bound to the ER synthesise extracellular proteins such as digestive enzymes and hormones (see page 82).

(f) Lysosomes

Lysosomes are fluid-filled sacs enclosed by a single membranous envelope which are found in most types of animal cells and probably also in plant cells. They are very variable in size and shape, possibly arising by fusion of small vesicles budded off from the edges of the Golgi apparatus or the ER. The high concentration of hydrolytic enzymes which they contain plays an important part in the normal intracellular digestive breakdown occurring during cell growth and repair. They are also involved in the digestion of pinocytosed and phagocytosed material entering the cell (see page 80) and in the cellular defences against attack by bacteria, viruses or toxic substances.

(g) Centrioles

Most animal cells and some plant cells contain a pair of centrioles which lies in the cytoplasm near to the nuclear membrane. A centriole is made of nine sets of triplet tubules arranged in a ring to make a hollow cylinder about 400 nm long and 200 nm in diameter (see diagram 2.7). The members of this centriole pair lie close together, often orientated at right angles to each other.

2.7   A centriole

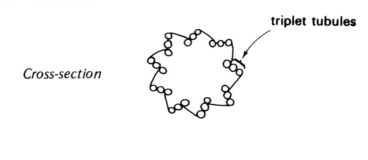

triplet tubules

Cross-section

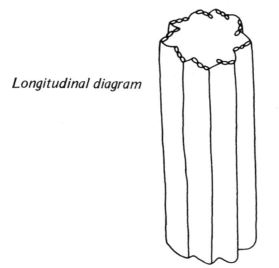

Longitudinal diagram

Centrioles are usually self-replicating and they have an important function during cell division when they separate and move to form the poles of the nuclear spindle (see page 92). The basal granules (kinetosomes) of cilia and flagella are structurally very similar to centrioles.

(h) Vacuoles

Vacuoles are fluid-filled droplets bounded by a unit membrane, the tonoplast. They are particularly conspicuous in plant cells.

There are many types of vacuoles:

food vacuoles	—	containing food materials
excretory vacuoles	—	containing waste materials
contractile vacuoles	—	involved in the maintenance of internal pressure in unicellular organisms
plant cell vacuoles	—	involved in the maintenance of the turgidity of plant tissues

A vacuole may fall into more than one of these categories.

The central vacuole of a mature plant cell may fill a large portion of the cell volume and restrict the cytoplasm to a thin layer inside the plasma membrane. The composition of the vacuole fluid is controlled by the selective permeability of the tonoplast membrane. A wide range of ingredients such as food materials, salts, waste substances and pigments occur in the fluid and may be involved in maintaining cell turgidity, often an important aspect of the support system of plant tissues.

(i) Granules

Cells may contain a variety of granules:
storage granules, e.g. starch, glycogen etc.
pigment granules
secretory granules of many types.

(j) Microtubules and microfilaments

Microtubules, hollow protein tubes 15–25 nm in diameter, are found in the cytoplasm in association with centrioles, cilia, flagella, sperm tails, nerve cell processes and other structures. Microfilaments, with a smaller diameter of 4–6 nm, also occur.

4 Nuclear components

If a cell is to continue its essential functions over a long period of time it must possess a nucleus. Some cells, such as striated muscle cells and fungal cells, are multinucleate but most contain a single nucleus.

(k) The nuclear membrane (nuclear envelope)

The nuclear membrane separates the nucleus from the surrounding cytoplasm. It is a flattened sac consisting of two layers of unit membrane which are pierced by small pores. The traffic of materials between the nucleus and the cytoplasm is controlled by this membrane.

(l) The nucleolus

A nucleolus is a dense spherical body of variable size which is *not* enclosed in a unit membrane. Several nucleoli may occur in one nucleus, each consisting largely of RNA and protein. The synthesis of ribosomal RNA takes place in the nucleoli, which are bigger in cells producing large amounts of protein. Special constricted regions on some of the chromosomes (nucleolar organisers) are associated with the nucleoli and are believed to be the site of nucleoli formation.

(m) Chromosomal material

Chromosomes are long, slender threads of DNA and histone protein which are usually dispersed throughout the nucleus and difficult to identify. They become very conspicuous, however, during cell division when the threads shorten and thicken by coiling until they become visible with the light microscope.

Generally the somatic cells of multicellular organisms possess a constant number of chromosomes, the diploid number, which is characteristic of the species. The chromosome complement is made up of two matching sets of chromosomes. The partners are called homologous pairs and they are similar in size and shape. But there are important exceptions to this generalisation. The gametes contain only half the usual chromosome number, the haploid number; algae and protozoans usually possess only a single, haploid set of chromosomes and some cells, particularly those of higher plants, contain various multiples of the haploid number such as triploid or tetraploid.

Functionally chromosomes are the carriers of the genes, those sections of the DNA strands which control RNA production and, ultimately, the protein synthesis in the cell (see page 82). By this means the genes on the chromosomes are the fundamental directors of cellular structure and function.

FURTHER READING

Electron Micrographs
Fawcett, D. W., *An Atlas of Fine Structure*, 1966, Saunders
Grimstone, A. V., *The Electron Microscope in Biology*, 1968, Arnold
Hurry, Stephen W., *The Microstructure of Cells*, 1965, Murray

General Texts
Scientific American, *The Living Cell*, 1965, Freeman
Swanson, C. P., *The Cell*, 3rd edition, 1969, Prentice Hall

Offprints
Scientific American articles may be obtained separately; for address see book list, page 117
de Duve, C., *The Lysosome*, May 1963
Fox, C. F., *The Structure of Cell Membranes*, Feb. 1972
Goodenough, V. W. and Levine, R. P., *The Genetic Activity of Chloroplasts and
    Mitochondria*, Nov. 1970
Green, D. E., *The Mitochondrion*, Jan. 1964
Margulis, L., *Symbiosis and Evolution*, Aug. 1971
Neutra, M. and Leblond, C. P., *The Golgi Apparatus*, Feb. 1969
Nomura, M., *Ribosomes*, Oct. 1969
Racker, E., *The Membrane of the Mitochondrion*, Feb. 1968
Wessells, N. K., *How Living Things Change Shape*, Oct. 1971

More Advanced Texts
Ambrose, E. J. and Easty, D. M., *Cell Biology*, 1970, Nelson
Loewy, A. G. and Siekevitz, P., *Cell Structure and Function*, 2nd edition, 1969 Holt,
    Rinehart and Winston
Novikoff, A. B. and Holtzman, E., *Cells and Organelles*, 1970, Holt, Rinehart and Winston
Oxford Biology Readers — short monographs by leading scientists, see biochemistry and
    cell-structure list

3 Cell Chemistry

Introduction

In order to understand how the cell functions it is essential to be familiar with the language of organic chemistry — the chemistry of the carbon compounds. While the cell makes use of many inorganic compounds, its functional complexities are made possible by the utilisation of organic macromolecules. These macromolecules employ the special ability of carbon atoms to link together into chain or ring structures, an ability which accounts for the almost unlimited number of carbon compounds and their wide variety of properties.

Carbohydrates, proteins, lipids and nucleic acids are the four groups of organic molecules most concerned in cellular structure and function. The large molecules in these classes are usually built from smaller repeating units linked together into long chains. Thus carbohydrate macromolecules are made up from small sugar units joined in a characteristic way, protein macromolecules are built from amino-acid units, and nucleic acids are constructed with three different building units, sugars, phosphate and organic bases. By such economical means living things can synthesise an enormous range of complex molecules to suit their diverse needs using only a relatively small number of basic components.

The carbohydrate group exists in a range of molecular sizes. Small monosaccharide sugars are the basic unit of the group. These are, most commonly, six-carbon molecules but five-carbon and three-carbon sugars are also important in living organisms. Monosaccharides form disaccharides by pairing together with a glycoside link and join up in larger numbers to produce the long-chain polysaccharide macromolecules. The monosaccharide glucose is the primary source of cellular energy. It is synthesised from inorganic molecules during photosynthesis and it is stored as the relatively insoluble polysaccharides, chiefly starch in plant cells and glycogen in animal cells. Other polysaccharides and carbohydrate derivatives, such as cellulose, lignin, pectin and chitin, act as important structural cellular components.

Table   3.1

Organic macromolecules — construction patterns

Group	Basic Unit	Linkage	Macromolecule
Carbohydrate	Monosacc-haride	Glycoside link	Polysaccharide (complex carbohydrate)
Protein	Amino-acid	Peptide link	Polypeptide (protein)
Nucleic Acid	Base Sugar Phosphate } Nucleotide	Sugar-phosphate link	Nucleic acid DNA or RNA

Proteins are made up of amino-acid units linked in large numbers to form long-chain macromolecules. They form part of many structural elements in the cell and they regulate cellular function by means of their enzymatic action (see page 55). An enormous variety of different proteins are required to catalyse all the biochemical reactions within the cell as enzymes are very specific. This diversity of protein macromolecules is provided by the wide range of possible permutations and combinations in which the twenty common amino-acid units may come together.

Lipids are a mixed group of chemically dissimilar substances united by the common properties of insolubility in water and solubility in certain organic solvents. Fatty acids are one of the basic units found in this group along with a variety of other components. Although lipid molecules may be quite large they do not reach macromolecular proportions. Lipids play a major part in cell membrane construction where their property of water insolubility is important. They may also act as reservoirs of stored energy and as insulating material.

The two nucleic acids, DNA and RNA, are large macromolecules built from three different types of molecular unit. Five-carbon sugars (pentoses), nitrogenous bases and phosphate are linked precisely together to make long chains. The arrangement of the bases on the nucleic acid chains provides a genetic information code by means of which the nucleic acids are the ultimate controllers of cellular function (see page 82).

Carbohydrates

Carbohydrates have three major roles to play in cells:
   1 Simple carbohydrates are the principal energy source in the cell.
   2 Long-chain carbohydrates act as food storage compounds.
   3 Long-chain carbohydrates form some of the structural components.

Carbohydrate classification

Carbohydrates are made up of the elements carbon, hydrogen and oxygen.
Hydrogen and oxygen are often present in the ratio of 2:1, as in water.
   The carbohydrate group is divided into three classes, according to the
number of carbon atoms which they contain.
   1 Monosaccharides — these are simple sugars containing less than 10
carbon atoms.
   2 Disaccharides — these are more complex carbohydrates formed by the
linkage of two monosaccharide units.
   3 Polysaccharides — these carbohydrates are macromolecules with high
molecular weights which are made of many monosaccharide units linked
together.

1 Monosaccharides

The monosaccharide group is important both because it contains many
biologically significant compounds and because these units are the basis of
the more complex carbohydrates.
   The group can be subdivided according to the number of carbon atoms in
the molecule:

trioses — C_3 sugars
pentoses — C_5 sugars
hexoses — C_6 sugars

   The group may also be subdivided into two series according to the linkages
between the carbon chain and the hydrogen and oxygen atoms. Variation in
linkage accounts for differing properties in the members of each series.
   (a) The aldose series has a terminal aldehyde group:

$$H - \underset{|}{C} = O$$
$$\underset{|}{C}$$

(b) The ketose series has a sub-terminal ketone group:

$$\begin{array}{c} C \\ | \\ C = O \\ | \\ C \end{array}$$

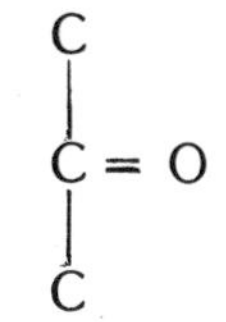

Two different structural forms, a chain or a ring, can often occur in the same monosaccharide (see diagram 3.1).

3.1   Monosaccharides

TRIOSE SUGARS
Structural isomers
Example – triose sugars, general formula $C_3H_6O_3$

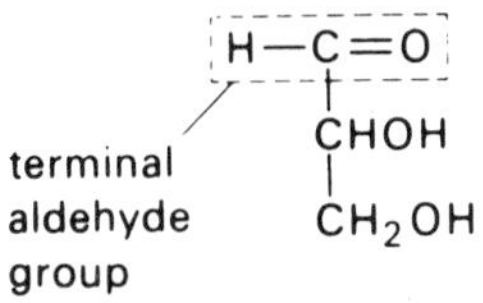

terminal aldehyde group

$$\begin{array}{c} H-C=O \\ | \\ CHOH \\ | \\ CH_2OH \end{array}$$

Glyceraldehyde
(an aldotriose)

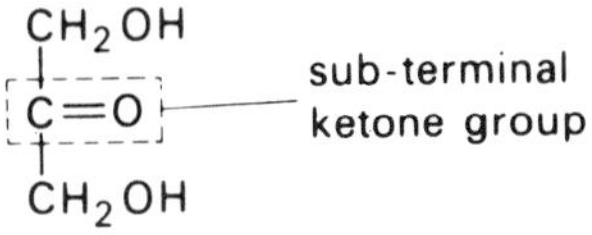

$$\begin{array}{c} CH_2OH \\ | \\ C=O \\ | \\ CH_2OH \end{array}$$ sub-terminal ketone group

Dihydroxyacetone
(a ketotriose)

Optical isomers
Example – glyceraldehyde

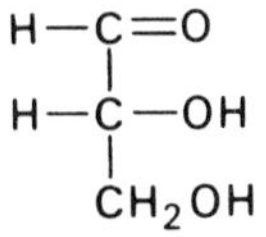

$$\begin{array}{c} H-C=O \\ | \\ H-C-OH \\ | \\ CH_2OH \end{array}$$

D (+) – glyceraldehyde
(+ = dextro-rotatory,
plane of polarised
light rotated to the right)

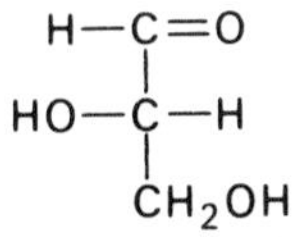

$$\begin{array}{c} H-C=O \\ | \\ HO-C-H \\ | \\ CH_2OH \end{array}$$

L (−) – glyceraldehyde
(− = laevo-rotatory,
plane of polarised
light rotated to the left)

PENTOSE SUGARS
Example – ribose

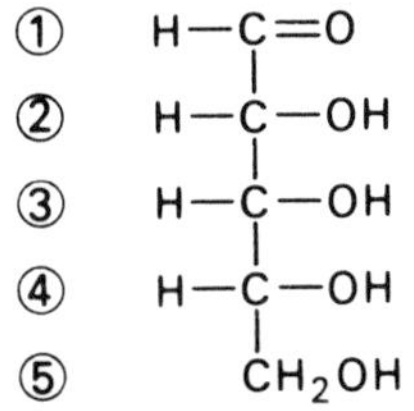

① $\quad H-C=O$
② $\quad H-C-OH$
③ $\quad H-C-OH$
④ $\quad H-C-OH$
⑤ $\quad CH_2OH$

ribose
chain structure

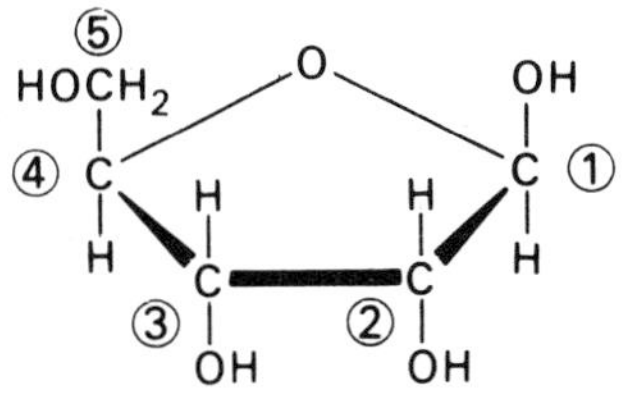

ribose
ring structure

Isomerism occurs commonly among carbohydrates and, as this phenomenon is often biologically significant, it will be discussed here. Structural isomers are compounds with identical atoms united by different bonds to form different molecular structures. Glyceraldehyde and dihydroxyacetone exemplify a structural isomer pair (see diagram 3.1). As the number of carbon atoms in a molecule increases so the number of possible isomers mounts.

Optical isomers can also occur when the spatial arrangement of the four different groups attached to a carbon atom varies (see diagram 3.1). Mirror-image pairs of molecules are produced which have opposite turning effects upon a beam of polarised light. Such small variations may be important in biochemistry as a cell often distinguishes optical isomers and can utilise only one member of an isomer pair.

Trioses — general formula $C_3H_6O_3$

Triose sugars occur as intermediate products along the photosynthetic and respiratory pathways, usually as triose phosphates (see page 39). The two isomers glyceraldehyde and dihydroxyacetone are an aldotriose and a keto-triose respectively.

Pentoses — general formula $C_5H_{10}O_5$

Pentose sugars are important constituents of the nucleic acids. Ribose is a sub-unit in RNA and its derivative deoxyribose, which lacks an oxygen atom, is found in DNA (see page 49). Ribose is also a component of the energy-rich compound adenosine triphosphate, ATP, and the nucleotide co-enzymes (see page 52). Two other pentoses, xylose and arabinose, are found in plant gums.

Hexoses — general formula $C_6H_{12}O_6$

Hexose sugars are common carbohydrates, forming the basic units of the disaccharides and many of the polysaccharides. Two hexoses, glucose and fructose, occur naturally; glucose is especially important because it is an immediate source of cellular energy.

(a) Glucose   This aldohexose sugar occurs very widely and it is the usual starting point of respiration. Glucose units are found in many complex carbohydrates.

(b) Fructose   This ketohexose sugar occurs naturally in fruit, nectar and honey. Fructose units are found in the polysaccharide inulin.

Table   3.2

MONOSACCHARIDES – the basic unit of the carbohydrates

Type of Monosaccharide	Aldoses	Ketoses
Trioses C_3 sugars	Glyceraldehyde	Dihydroxyacetone
Pentoses C_5 sugars	Xylose Arabinose Ribose	
Hexoses C_6 sugars	Glucose Galactose Mannose	Fructose

This list includes only the common monosaccharides; at least 20 are known.

POLYMERISED CARBOHYDRATES – formed by condensation of monosaccharide units

Monosaccharide unit	Common Disaccharides	Common Polysaccharides
1. Pentose units		
Xylose		Xylan – condensation product of xylose
Arabinose		Araban – condensation product of arabinose
2. Hexose units		
Glucose only	Maltose – condensation product of 2 glucose units	Cellulose – condensation chain of 300 – 3000 glucose units
		Starch – condensation chain of approx. 300 glucose units, linkage diff. from cellulose
		Glycogen – branched condensation chain of glucose units
Fructose only		Inulin – condensation product of fructose units
Glucose and fructose	Sucrose – condensation product of one glucose unit and one fructose unit	
Glucose and galactose	Lactose – condensation product of one glucose unit and one galactose unit	

3.2   *Monosaccharides – hexose sugars*

Chain structures

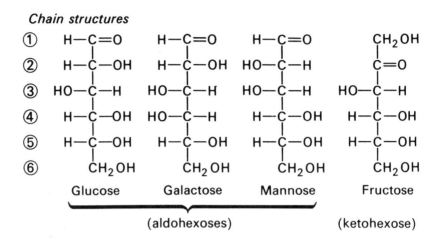

Ring structures
Example – the glucose ring

α-glucose   β-glucose

These two ring forms are distinguished by the arrangement of the H and OH groups attached to carbon atom number 1

(c) Galactose   This is not a natural sugar but galactose units make the disaccharide lactose.

(d) Mannose   Mannose occurs only as a component of complex carbo-hydrates such as some mucilages.

2 Disaccharides

Disaccharide sugars are formed by the condensation of two monosaccharide hexose molecules. This is a dehydration synthesis which joins the hexose units by a glycoside link and eliminates a molecule of water (see diagram 3.3). The hydrolysis of a disaccharide breaks the molecule into its monosaccharide components with the addition of water; it is a reversal of the process of condensation. The hexose units of the long-chain carbohydrates are usually in the ring form and the glycoside links are of different types (α and β) depending upon which carbon atoms in the sub-units they join together.

(a) Maltose   A condensation product of two glucose molecules joined by a 1,4 α link (see diagram 3.3). Maltose occurs in soya beans and germinating cereals. It is formed when the polysaccharide starch is hydrolysed by the enzyme amylase.

(b) Lactose   This sugar, which is present in all mammalian milk, is the condensation product of one glucose molecule and one galactose molecule joined by a 1,4 β link. It is not found in plants.

(c) Sucrose   This is the most common of all sugars as it is widely found in plants. It is the condensation product of glucose and fructose (with a 1,5 α link) and, unlike the other sugars, it is *not* a reducing agent and may there-fore easily be distinguished from them. Polysaccharide storage carbohydrates in plants are hydrolysed to sucrose for transport since it is quite soluble. Sucrose is extracted commercially from sugar-cane and sugar-beet.

3 Polysaccharides

The polysaccharide group includes many compounds which form structural components or storage products in plants and animals. These long-chain molecules are formed by the condensation of large numbers of monosaccharide units. Their properties vary with their constituent units, their type of linkage and the spatial arrangement of their chains (see diagram 3.4).

3.3   *Disaccharides (two rings linked together)*

1 *Maltose — two glucose sub-units joined by a $1\underset{\alpha}{-}4$ glycoside link*

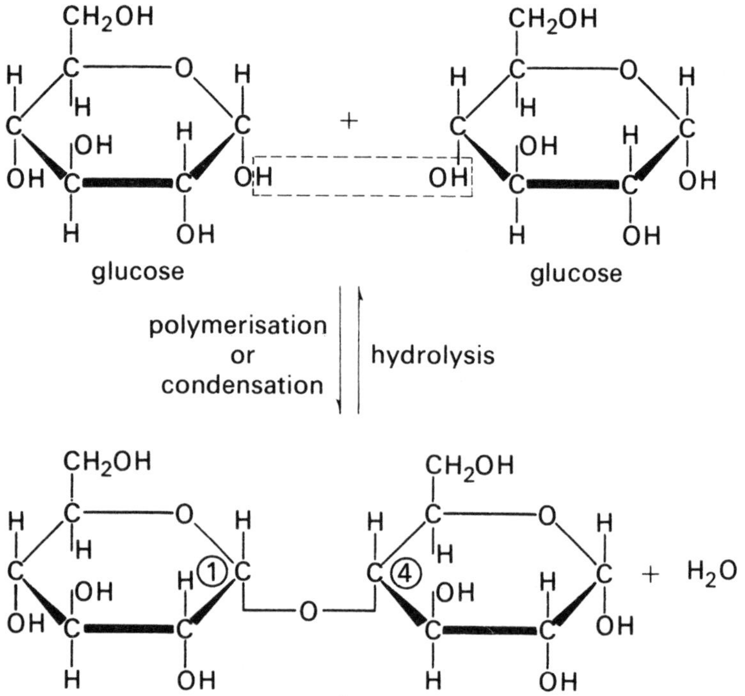

glucose   glucose

polymerisation
or   hydrolysis
condensation

maltose $+\ H_2O$

2 *Lactose — galactose and glucose sub-units joined by a $1\underset{\beta}{-}4$ glycoside link*

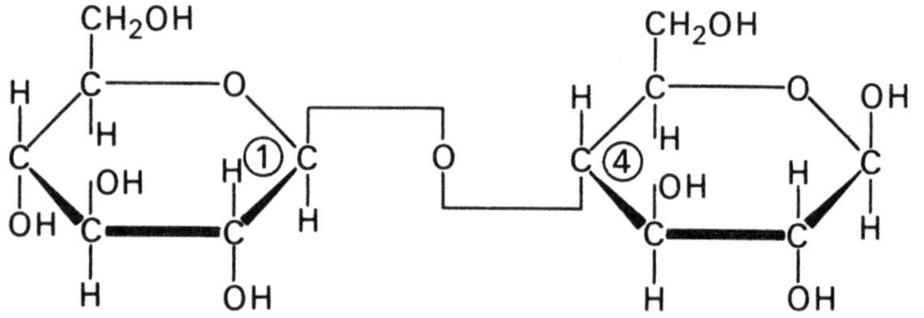

galactose   glucose

3 *Sucrose — fructose and glucose sub-units joined by a $1\underset{\alpha}{-}5$ glycoside link*

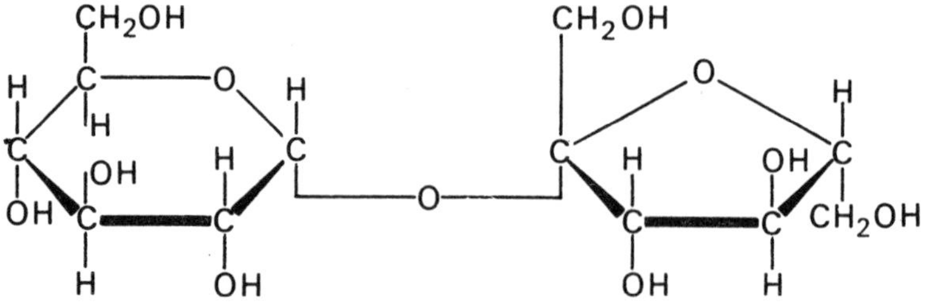

glucose   fructose

Pentosans

Pentosans are condensation products of pentose (C_5) sugars. The group includes xylan, which is found in woody, lignified plant tissues, and araban, which occurs in plant gums.

Hexosans

Hexosans are the more important of the two polysaccharide groups and they are formed by condensation of hexose (C_6) sugars. Hexosans are either insoluble in water or they make colloidal solutions. Most can be identified by the formation of coloured products with iodine.

Structural hexosans
(a) Cellulose   A condensation product of glucose (with 1,4β links). Cellulose is the main structural component of plant cell walls where its long unbranched molecules lie parallel to give great strength (e.g. cotton).

Storage hexosans
(b) Starch   Another type of long-chain condensation product of glucose (with 1,4 α links). Starch occurs in several forms in most *plants*, where it acts as the main reserve food supply.
(c) Glycogen   Another glucose condensation product with branched chains and both 1,4 and 1,6 α links. It acts as a carbohydrate store in *animal* tissues such as liver and muscle (see diagram 3.4).
(d) Inulin   A condensation product of fructose which acts as a food reserve in a few plants, notably dahlias, dandelions, chicory and Jerusalem artichokes.

4 Carbohydrate derivatives

Heterosaccharides

Substances in this category are closely related to carbohydrates. They include some of the structural materials such as chitin, a component of the invertebrate exoskeleton, and the derivatives commonly found in plants such as lignin and pectin. The mucopolysaccharides of animal tissues such as mucus, cartilage, tendons and intercellular cement are also carbohydrate derivatives.

3.4   *Polysaccharides (chains of rings)*

1 SIMPLE CHAINS

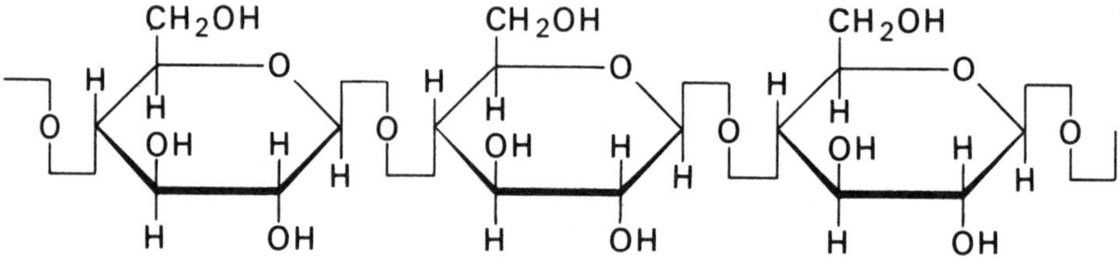

a) *Cellulose — glucose units joined by a 1$\underset{=}{\beta}$4 link*

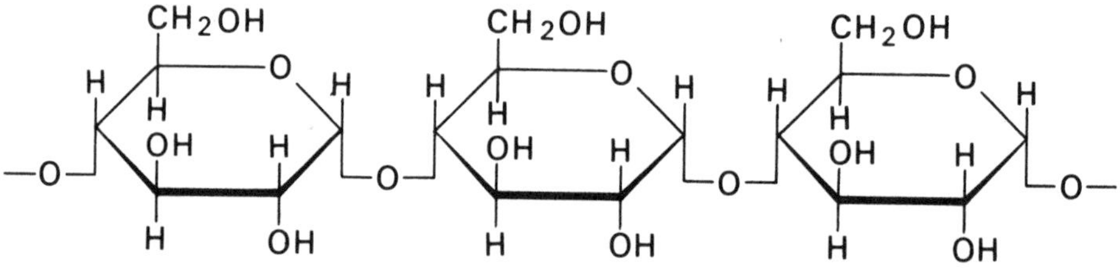

b) *Starch — glucose units joined by a 1$\underset{=}{\alpha}$4 link*

2 BRANCHED CHAINS

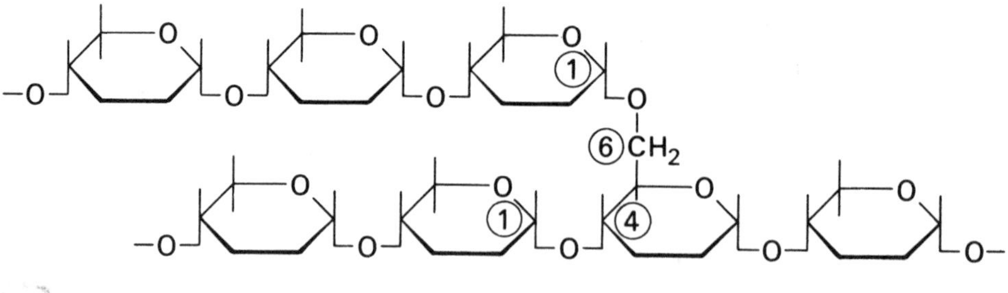

Glycogen — chains of glucose units joined by 1$\underset{=}{\alpha}$4 links and cross-linked by 1$\underset{=}{\alpha}$6 glycoside links

Phosphorylated carbohydrates

Another important group of carbohydrate derivatives are the phosphorylated monosaccharides. These play a major part in cellular reactions as they are very much more reactive than their unmodified parent molecules. Glucose-6-phosphate is the phosphorylated derivative of glucose where the H-atom attached to carbon atom number 6 has been replaced by a phosphate group. Fructose may also occur in a phosphorylated form (see diagram 3.5). Both these compounds are intermediaries in the respiratory pathway (see page 69).

3.5 *Phosphorylated monosaccharides*

1 *Glucose–6–phosphate*

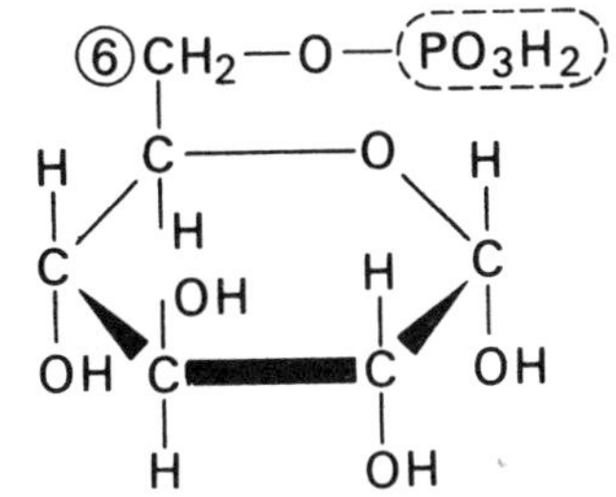

(P) = phosphate = H_2PO_3 group

2 *Fructose–6–phosphate*

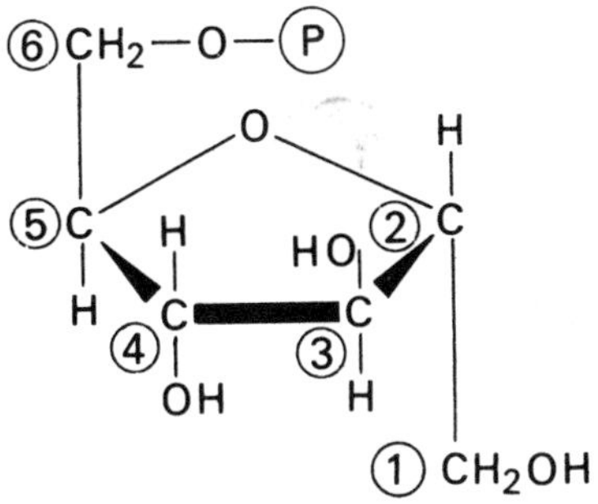

3 *Fructose 1,6 diphosphate*

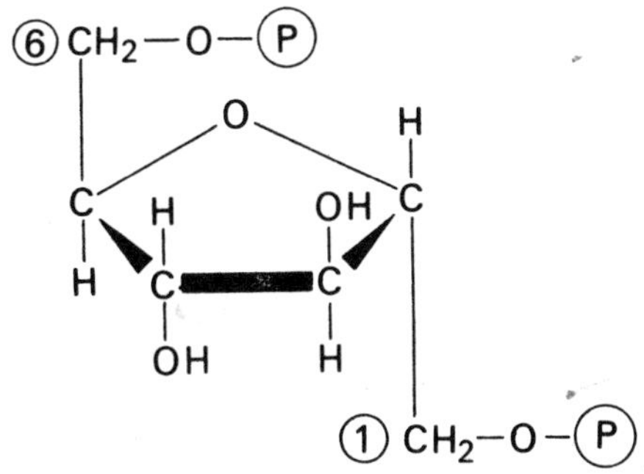

Proteins

Proteins play many important roles in the cell:
  1 The fibrous proteins act as building materials for the molecular 'scaffolding' of the cell structure.
  2 The globular and conjugated proteins are of vital importance to cellular function since many of them act as reaction-catalysing enzymes. Others perform specific tasks such as the carriage of oxygen, the defence of the animal (antibodies), the transfer of chemical messages (hormones), the clotting of blood and many more.
  3 Proteins may act as an energy source.

Protein structure

Proteins are very large organic macromolecules. They are nitrogenous compounds containing carbon, hydrogen, oxygen and nitrogen. Phosphorus and sulphur are often also found in proteins.
  Protein macromolecules are polymers formed by the condensation of molecular units called amino-acids. All amino-acids have the general structure:

$$H_2N-\underset{\underset{R}{|}}{\overset{\overset{H}{|}}{C}}-COOH$$

NH_2 = the amino group

$COOH$ = the carboxyl group

R = a variable atomic group which is different in each amino-acid

About twenty amino-acids occur naturally in living cells (see diagram 3.6). Optical isomers are found among them as the central carbon atom is attached to four different groups (except in glycine) and so is asymmetrical. All the amino-acids that have been found incorporated into proteins are of the L-variety (see diagram 3.1).
  Many hundreds or thousands of amino-acid units are joined together to form a protein chain. The amino group of one acid fastens to the carboxyl group of its neighbour, a molecule of water is eliminated and the 'peptide link' (—NH—CO—) is established (see diagram 3.8). A dipeptide is produced when two amino-acids are linked together by peptide bonds and a polypeptide is the result of the polymerisation of many amino-acid units.
  Protein structure involves not only the primary consideration of chemical

3.6   *Proteins — amino-acids*

General formula

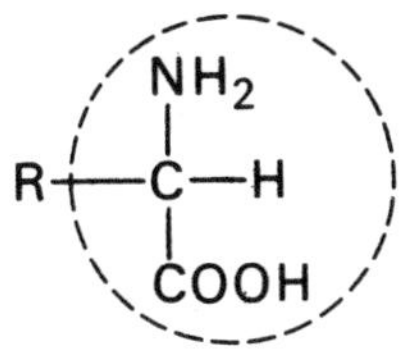

NH₂
|
R—C—H
|
COOH

Example — glycine

NH₂
|
H—C—H
|
COOH

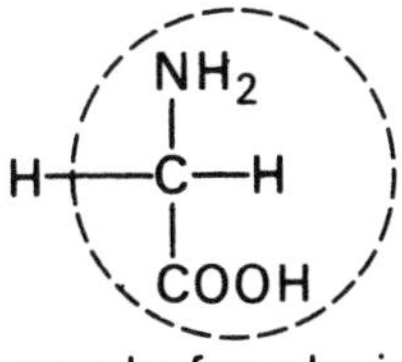

Formula for glycine
where R = H

Other common amino-acids

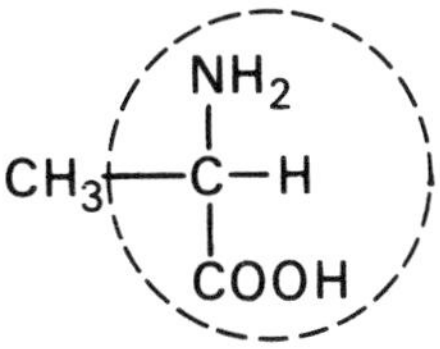

CH_3—C—H with NH₂ and COOH — **Alanine**

HO—CH_2—C—H with NH₂ and COOH — **Serine**

HS—CH_2—C—H with NH₂ and COOH — **Cysteine**

$HOOC$—CH_2—C—H with NH₂ and COOH — **Aspartic acid**

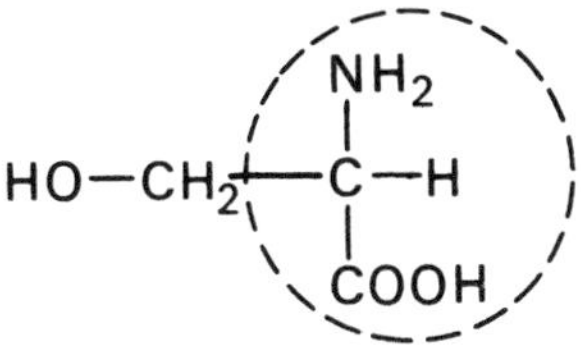

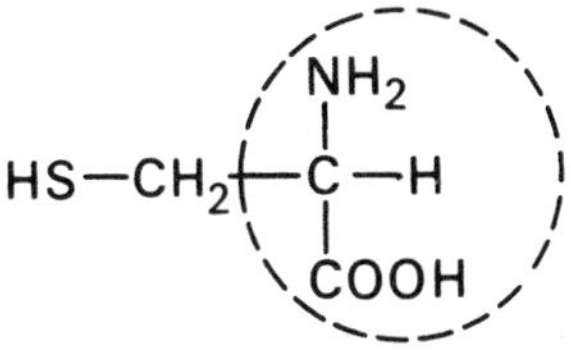

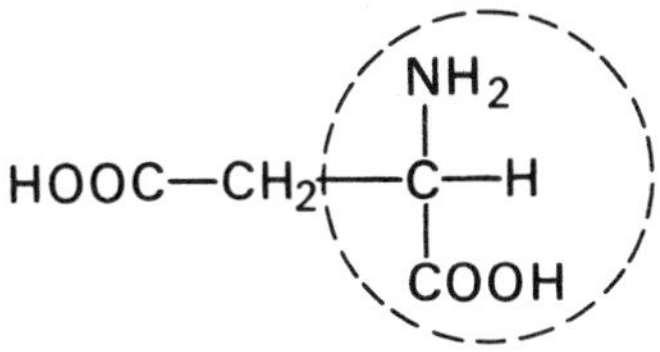

$HOOC$—CH_2—CH_2—C—H with NH₂ and COOH — **Glutamic acid**

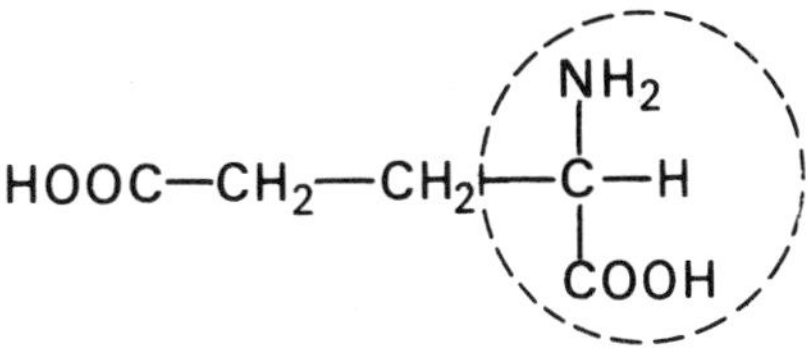

CH_3, CH, CH_3—C—H with NH₂ and COOH — **Valine**

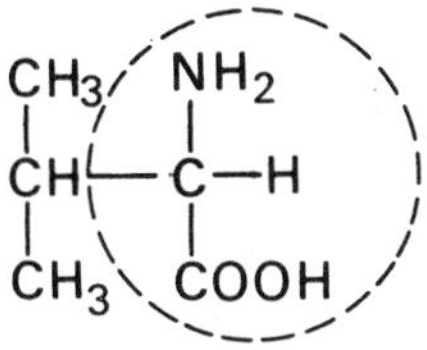

CH_3, CH, CH_3—CH_2—C—H with NH₂ and COOH — **Leucine**

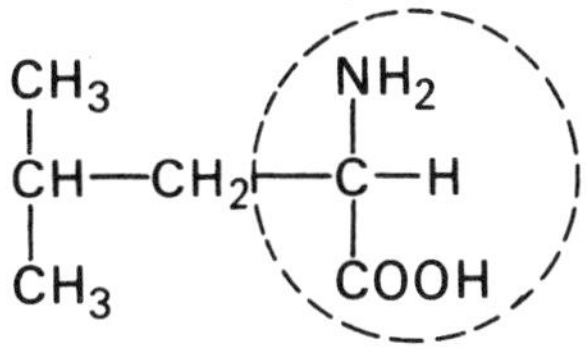

There are about twenty common amino-acids

composition but also the important secondary consideration of molecular shape.

The *primary* structure of a protein molecule is concerned with the specific sequence of the particular amino-acid units which make up that molecule. Since a protein may contain any or all of the 20 different amino-acids and these may be linked in any sequence and in any quantity, the scope for variation in primary protein structure is enormous.

The *secondary* structure of a protein relates to the way in which the molecule is arranged in space. The polypeptide chain is twisted and folded to produce a stable, three-dimensional structure which may be maintained by and of three types of bonds. These bonds (see diagram 3.7) are:

1 The *ionic* bond which forms when two oppositely charged groups are attracted electrostatically.

2 The *hydrogen* bond which develops when two groups are linked through a shared hydrogen atom.

3 The *disulphide* bond which arises between two amino-acids containing sulphur, such as cysteine. This is the strongest of the three bonds.

Protein classification

The protein group is divided into two categories — the simple proteins which are made up solely of protein and the conjugated proteins which contain a non-protein fraction.

A. The simple proteins

Since the properties of proteins are determined, in part, by molecular shape this factor is used in protein classification. On this basis there are two groups of simple proteins:

1 Fibrous proteins have extended thread-like polypeptide chains, which may be straight or twisted into a coiled spring held by hydrogen bonds between successive coils (see diagram 3.8). These molecules are strong and insoluble with resistance to changes in acidity and temperature.

The group includes the structural proteins such as collagen in skin, keratin in hair, nails and wool, myosin in muscle and elastin of cartilage and arteries.

2 Globular proteins possess polypeptide chains which are twisted into irregular, complex structures and held in place by the three types of chemical bond (described earlier). These molecules are chemically reactive, soluble and

3.7 Types of protein bond

The peptide bond

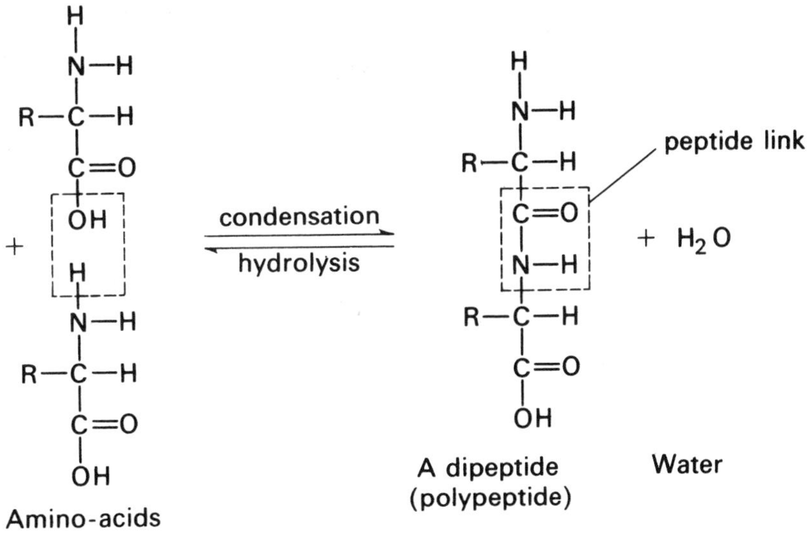

H
|
N—H
|
R—C—H
|
C=O
|
OH

+

H
|
N—H
|
R—C—H
|
C=O
|
OH

Amino-acids

$\xrightarrow[\text{hydrolysis}]{\text{condensation}}$

H
|
N—H
|
R—C—H
|
C=O
|
N—H
|
R—C—H
|
C=O
|
OH

peptide link

A dipeptide
(polypeptide)

+ H_2O

Water

Bonds between polypeptide chains

1. The ionic bond, an electro-static bond, links charged side groups of the poly-peptide chains.

2. The hydrogen bond links two side groups of the poly-peptide chains through a shared hydrogen atom. It is a weak bond.

3. The disulphide bond is a covalent bond linking sulphur-containing amino acids (e.g. cysteine) in adjacent polypeptide chains.

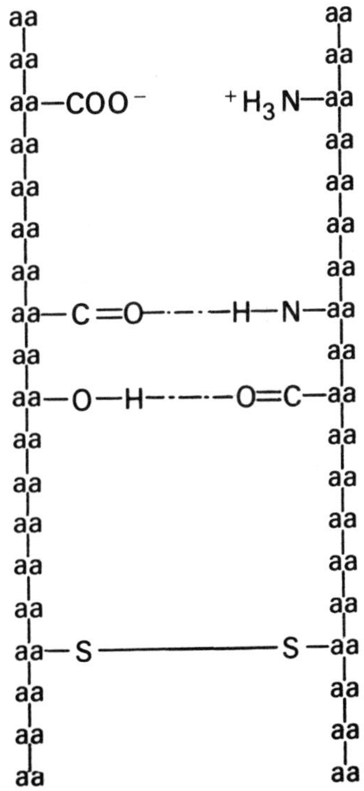

aa
aa
aa—COO⁻   ⁺H_3N—aa
aa
aa
aa
aa
aa—C=O—·—·—H—N—aa
aa
aa—O—H—·—·—O=C—aa
aa
aa
aa
aa
aa
aa—S————————S—aa
aa
aa
aa

3.8 *Protein structure*

PRIMARY PROTEIN STRUCTURE

Primary structure relates to the chemical composition of the protein and the amino acid sequence of its molecule.

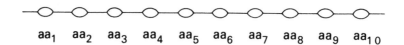

aa_1 aa_2 aa_3 aa_4 aa_5 aa_6 aa_7 aa_8 aa_9 aa_{10}

SECONDARY PROTEIN STRUCTURE

Secondary structure relates to the three-dimensional structure of the protein molecule.

Regular secondary protein structure

a) Straight chain – the polypeptide chain is straight or extended and may be joined to form sheets by hydrogen bonds between chains.

b) Helix – the polypeptide chain is twisted into a helical structure and held in position by hydrogen bonds between each amino-acid unit and its neighbouring amino-acid positioned three units away.

The <u>fibrous</u> proteins possess a regular secondary structure.

*Irregular secondary protein structure

Irregular secondary structure is found in the <u>globular</u> proteins. This type of structure may be very complex, involving helical and straight chain portions and various irregular secondary bonds, particularly disulphide bridges.

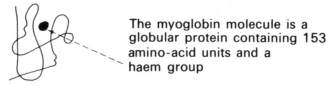

The myoglobin molecule is a globular protein containing 153 amino-acid units and a haem group

* This irregular three-dimensional configuration is sometimes known as the tertiary structure of the protein.

easily denatured (changed irreversibly) by temperature increase, pH change and some chemicals. Functional proteins, such as enzymes and hormones, are globular proteins.

B. The conjugated proteins

Some proteins are complex molecules in which globular proteins are combined with non-protein components. These are usually classified with reference to the nature of the non-protein fraction.

Lipoproteins possess a lipid derivative bound to the protein molecule. They are the major components of cell membranes.

Mucoproteins or glycoproteins are conjugates of proteins and carbohydrate derivatives. They are found intracellularly and in the blood plasma.

Nucleoproteins link the proteins histone or protamine with nucleic acids. Nucleic acids are often found associated with proteins in this way.

Chromoproteins combine a protein fraction with a pigmented molecular unit. Examples are haemoglobin and flavoprotein.

Metalloproteins combine protein molecules with metal atoms. Haemoglobin also falls in this category.

Phosphoproteins are conjugates of proteins with a phosphate group.

Lipids

Fats and their derivatives, together known as lipids, have many useful functions in the cell. They:

    1 act as storage compounds in animals, in the fruits and seeds of plants and in other organisms

    2 act as structural cellular components, particularily in cell membranes

    3 provide a rich source of energy, yielding twice as many calories, weight for weight, as carbohydrates or proteins

    4 provide electrical and thermal insulation.

Lipid structure

Lipids are a heterogeneous group. In general they contain the elements carbon, hydrogen and oxygen but they include proportionately less oxygen than the carbohydrates. Most lipids are esters, compounds formed by condensation between an alcohol and an acid. All lipids are insoluble in water but soluble in certain organic solvents.

Fatty acids are the acids found in many lipids. These are quite simple organic molecules made of long hydrocarbon chains which carry a terminal carboxyl group (—COOH) (see diagram 3.9). Acetic acid, CH_3COOH, is a simple fatty acid and stearic acid, $CH_3(CH_2)_{16}COOH$, is a more complex one. The terminal carboxyl group is polar and is water-soluble while the hydrocarbon group is non-polar and insoluble. Fatty acids therefore become less and less water soluble as the length of their hydrocarbon chain increases. Unsaturated fatty acids have one or more double bonds in their molecules whereas the maximum possible number of hydrogen atoms are incorporated into saturated fatty acids.

3.9 Lipids

Fatty acids

Examples — acetic acid, butyric acid

Fats and oils

Example — tristearin, a fat

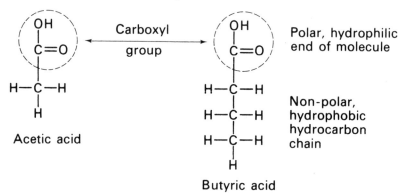

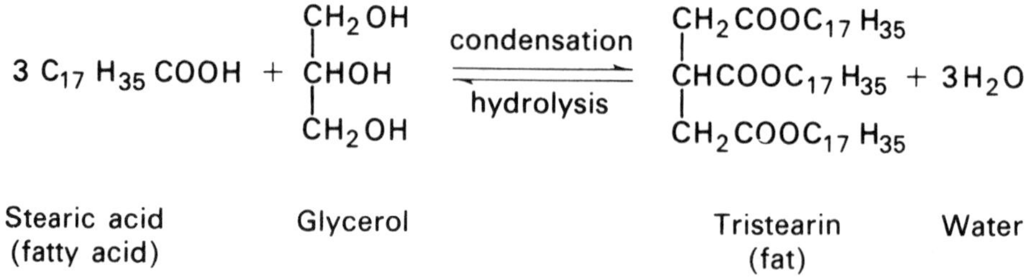

| Stearic acid (fatty acid) | Glycerol | | Tristearin (fat) | Water |

1 Fats and oils (neutral lipids)

Fats and oils are esters derived from glycerol, an alcohol with the formula $C_3H_5(OH)_3$, and fatty acids (see diagram 3.9). Fats are made largely of

3.10   *Lipids*

Phospholipids

Example – lecithin

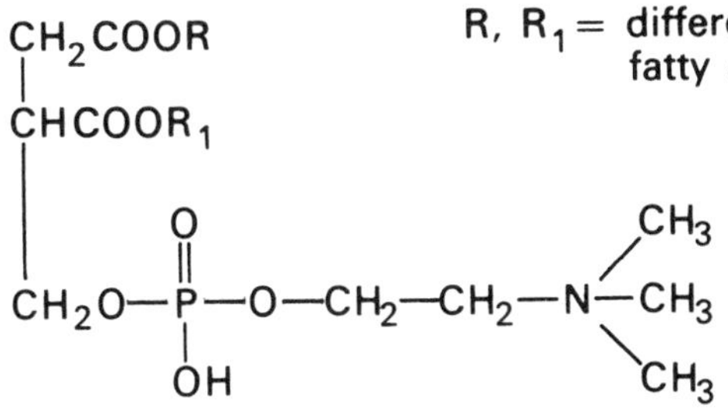

CH₂COOR   R, R₁ = different
| fatty acids
CHCOOR₁

Lecithin

Steroids

The steroid skeleton is made up of 17 carbon atoms, joined as four linked carbon rings

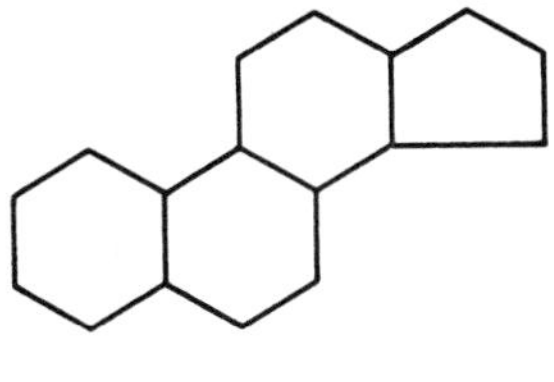

Example – testosterone, a male sex hormone

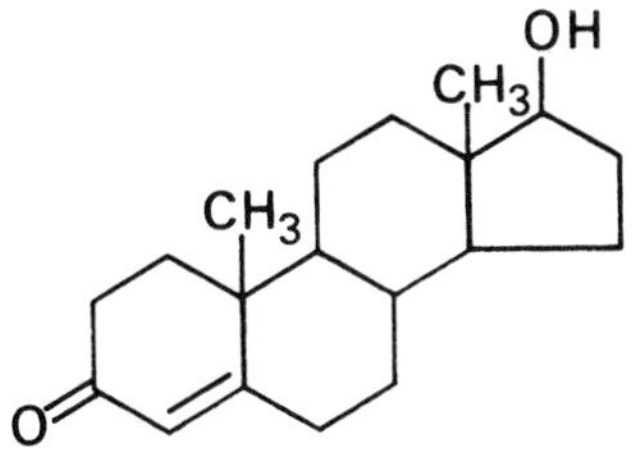

saturated acids and are solid at normal temperatures while oils include many unsaturated acids and are liquid at room temperature. These substances are deposited as food reserves in the storage organs of some plant tissues and in the adipose tissues of animals where the fat may also act as a heat insulator and as protective packing for delicate organs. Glycolipids are neutral lipids with carbohydrate side chains found in many cellular membranes.

2 Waxes

Waxes are esters of long chain fatty acids with long chain alcohols. Waxes serve to waterproof the outer surfaces of plant and animal tissues.

3 Phospholipids

The major phospholipids are esters of glycerol and a mixture of fatty acids and phosphoric acid. Phospholipids usually form structural components of cells. Lecithin is a constituent of cellular membranes and of egg yolk (see diagram 3.10). Cephalin occurs in the brain and acts as insulation material for nerves.

4 Steroids

Steroids are not true lipids but are derived from them. The sex hormones, cholesterol and the bile salts all belong in this category (see diagram 3.10).

Nucleic acids

The nucleic acids are of fundamental importance to the cell as they control all the cellular activities, directly and indirectly, by means of the synthesis of the cell proteins.

Both types of nucleic acid, deoxyribonucleic acid (DNA) and ribonucleic acid (RNA), are normally present in the nucleus of a cell. DNA seldom occurs outside the nucleus (it has recently been found in mitochondria and chloroplasts) whereas RNA is found in the nucleoli of the nucleus and in the cytoplasm, particularily in association with the ribosomes.

Nucleic acid structure

The nucleic acids are macromolecules built from three types of repeating sub-units:
    1 nitrogenous bases — purines and pyrimidines
    2 pentose sugars
    3 phosphate units.

1 Purine and pyrimidine bases

These organic bases belong to a series of related ring compounds in which the rings contain both carbon and nitrogen atoms. The purines adenine and guanine are made up of two interconnecting rings while the pyrimidines — cytosine, thymine and uracil — possess a single ring structure (for detailed structures see diagram 3.11).

2 Pentose sugar units

The pentose (C_5) sugars found in the nucleic acids are of two types, ribose and deoxyribose. They differ only in that ribose sugar contains an additional oxygen atom (see diagram 3.11).

3 Phosphate units

The phosphate unit acts as a link between one sugar group and the next (see diagram 3.11)
    In nucleic acids these three types of sub-unit are linked together in a specific way to make a larger unit, a nucleotide (see diagram 3.12). Many nucleotides are then joined by condensation to form a long chain which is the backbone of nucleic acid structure. This backbone is actually made up of alternating sugar and phosphate units, the bases are attached to the sugar units.

Ribonucleic acid

The RNA molecule is made up of long chains of nucleotides incorporating the pentose sugar ribose and any of the four bases — adenine and guanine (purines) or cytosine and uracil (pyrimidines). The base uracil is found exclusively in RNA while the other bases also occur in DNA. RNA chains may exist as long strands or as ring structures.

3.11   *Nucleic acid components*

Purines and pyrimidines (nitrogenous bases)

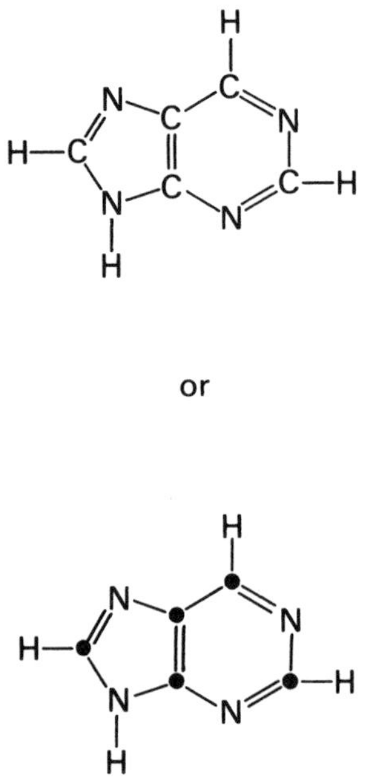

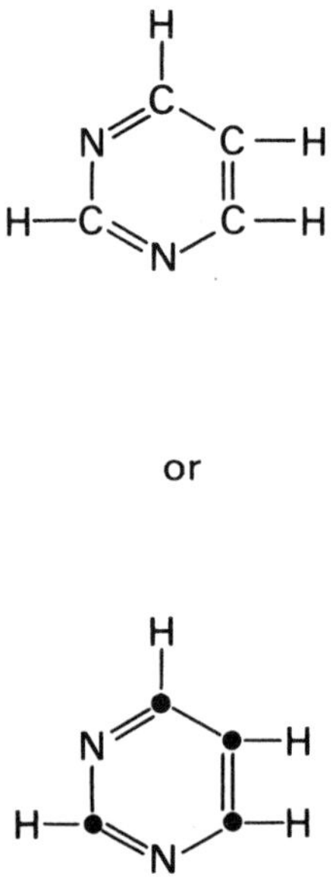

or

or

Purine skeleton–
a <u>double</u> carbon
and nitrogen ring

Pyrimidine skeleton–
a <u>single</u> carbon
and nitrogen ring

Purines – adenine and guanine

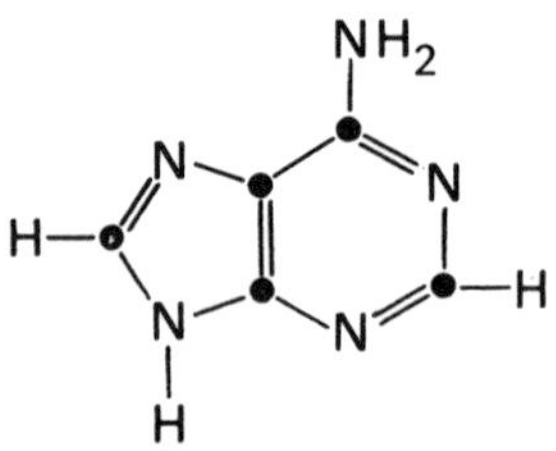

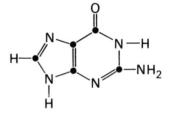

Adenine

Guanine

Pyrimidines — cytosine, thymine and uracil

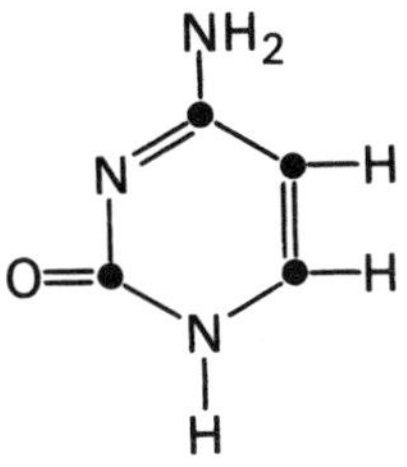

Cytosine 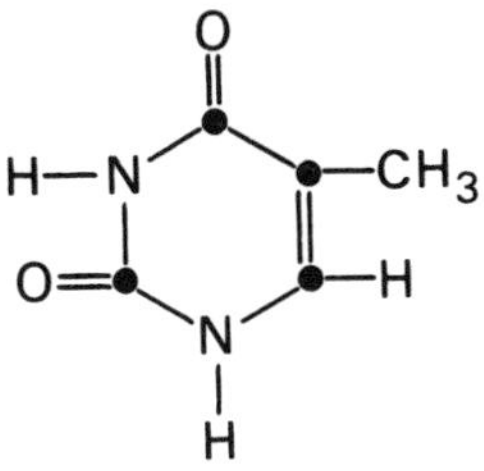   Thymine 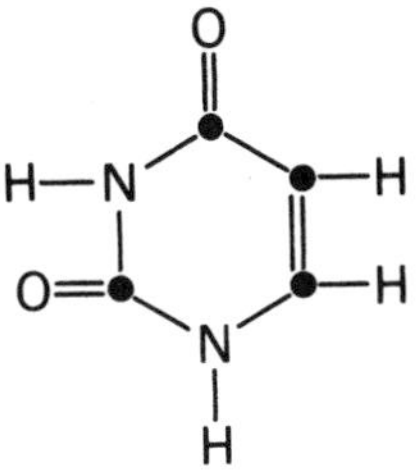   Uracil

Pentose sugar units

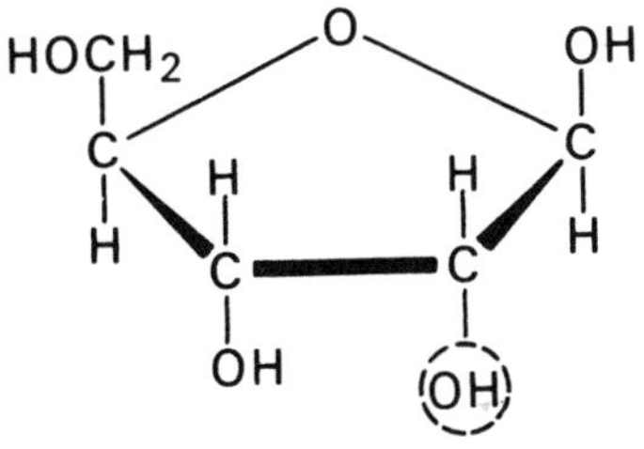

Ribose 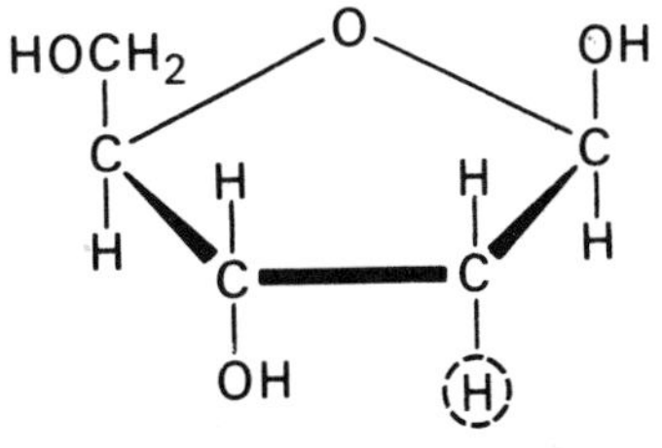   Deoxyribose

The phosphate group

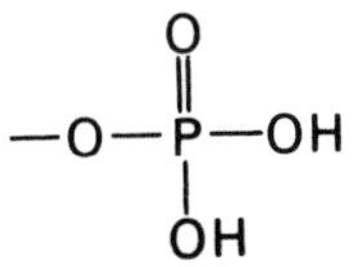

usually written $-\text{P}$

Deoxyribonucleic acid

DNA has a very high molecular weight and its structure is more complex than that of RNA. DNA is similarly made up of long chains of nucleotides but, in this case, the sugar unit is the pentose deoxyribose and the pyrimidine base uracil is replaced by thymine. DNA chains do not occur singly but pair together by weak hydrogen bonding across the bases to form a ladder-like construction. The bases can only link in a specific way; purine must always link with pyrimidine and, even more precisely, because of their molecular configuration, adenine must bond with thymine and guanine links with cytosine (see diagram 3.13).

The ladder-like molecule of DNA is twisted into the shape of a double helix. Its structure resembles a spiral staircase with bannisters made of alternating sugar/phosphate chains and steps made of bonded base pairs (this is the Watson/Crick model of DNA structure, see diagram 3.12).

Phosphorylated Nucleotides — high-energy compounds

The addition of a phosphate group to an organic molecule has the effect of greatly increasing the reactivity of that molecule — a property that has already been mentioned in relation to the phosphorylated monosaccharides (see page 39). Nucleotides may also acquire additional phosphate groups with a consequent increase in reactivity. This may be exemplified by adenine ribonucleoside triphosphate (adenosine triphosphate or ATP) which possesses two extra phosphate groups linked to the molecule by especially 'high-energy' bonds. This important molecule acts as an energy carrier during cellular reactions (see page 60).

Other important nucleotides

Many nucleotide derivatives function as co-enzymes (see page 59) acting in conjunction with specific enzymes as carrier molecules by transferring a group of atoms from one molecule to another. Nicotinamide adenine dinucleotide (NAD), its phosphorylated derivative NADP and flavin adenine dinucleotide (FAD) all act as hydrogen carriers (see page 64). Co-enzyme A, also a nucleotide derivative, acts as an acetyl carrier molecule (see page 72).

3.12   *Nucleic acids*

The nucleotide chain

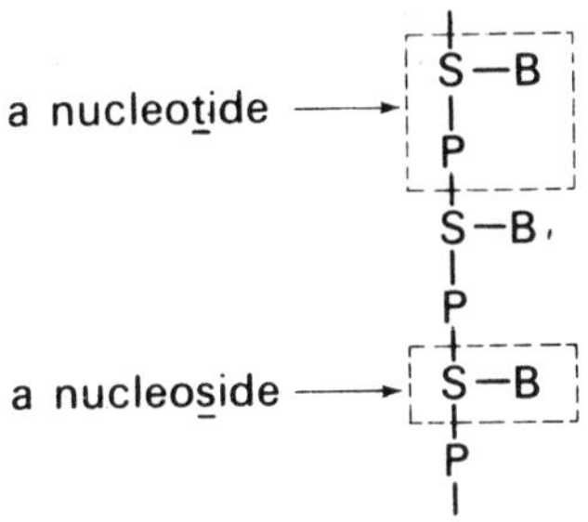

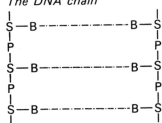

a nucleotide ——→

```
   +----+
   | S—B |
   |  |  |
   |  P  |
   +----+
```

a nucleoside ——→

```
   S—B,
    |
    P
   +------+
   | S—B  |
   +------+
    |
    P
    |
```

where —

S = sugar unit

P = phosphate unit

B = base unit

The RNA chain

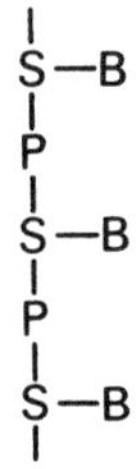

```
 |
 S—B
 |
 P
 |
 S—B
 |
 P
 |
 S—B
 |
```

where

S = <u>ribose</u> sugar

P = phosphate unit

B = any one of these four bases —
    adenine
    guanine
    <u>uracil</u>
    cytosine

The DNA chain

```
 |                    |
 S—B-----------B—S
 |                    |
 P                    P
 |                    |
 S—B-----------B—S
 |                    |
 P                    P
 |                    |
 S—B-----------B—S
 |                    |
```

where —

S = <u>deoxyribose</u> sugar

P = phosphate unit

B = any one of these four bases —
    adenine
    guanine
    <u>thymine</u>
    cytosine

Hydrogen bonds link the bases so that a purine base always links with a pyrimidine base —
adenine links with thymine   and   guanine links with cytosine

NUCLEIC ACIDS

DNA structure

The double chain of DNA is spiraled to make a <u>double helix</u>. Each component strand is made up of a — S — P — S — P — S — P — S — P — chain and the cross links represent purine-pyrimidine base pairs held together by hydrogen bonds. There are 10 cross links in each turn of the helix.

3.13   *Nucleic acids — the base pairs (purine-pyrimidine links)*

Adenine—thymine pair, linked by two *hydrogen bonds*

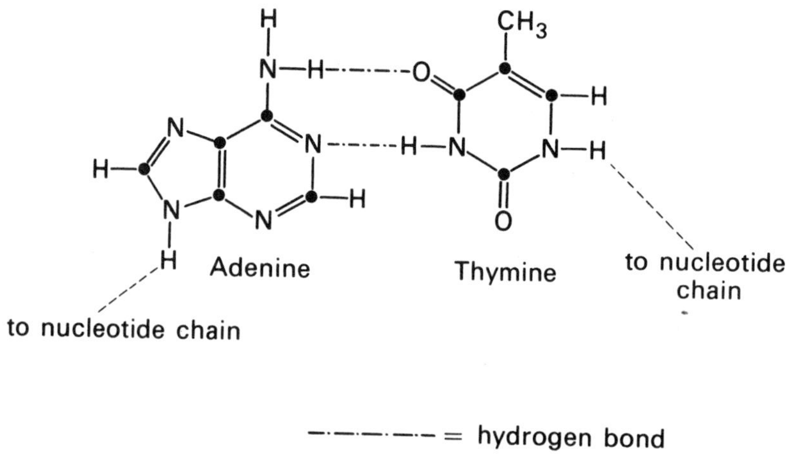

Adenine   Thymine   to nucleotide
                                                  chain

to nucleotide chain

------- = hydrogen bond

Guanine—cytosine pair, linked by three *hydrogen bonds*

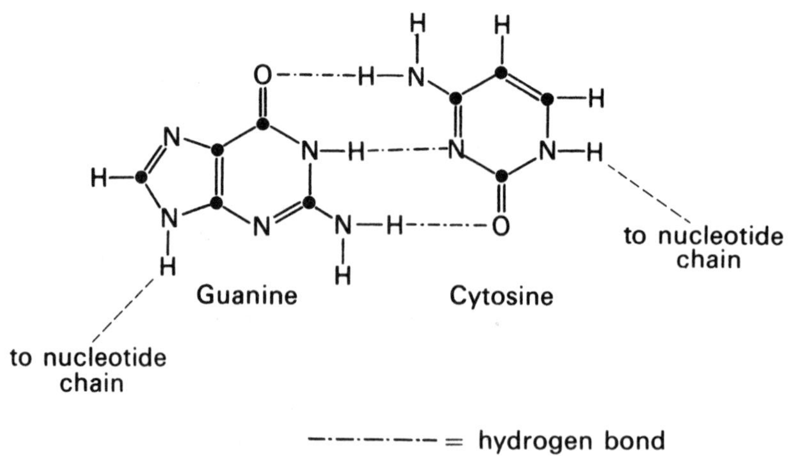

Guanine   Cytosine

to nucleotide   to nucleotide
   chain   chain

------- = hydrogen bond

Enzymes

Activities taking place in the cell are largely the product of innumerable chemical reactions of enormous variety and complexity. It is often essential that these reactions should occur very rapidly and in precise sequence. In a chemical laboratory reactions may be hastened by an increase of temperature or pressure or by an alteration in acidity; such measures are not possible under physiological conditions without causing damage to the fabric of the cell. An alternative method for speeding up reactions is the use of catalysts, substances which increase the rate of a reaction while remaining unchanged themselves. Catalysis is common in chemistry and it is thought to be practically universal among biochemical reactions.

Biochemical catalysts are called *enzymes*. These are always globular proteins with complicated three-dimensional structures. An enzyme is often associated with a non-protein component, a co-factor, which may be so tightly bound to its enzyme that a conjugated protein is formed.

Enzyme action

Enzyme action is very specific; each enzyme usually catalyses only one particular type of reaction. This specificity is reflected in the modern system of nomenclature where the enzyme name denotes the nature of the reaction that it catalyses. For example the suffix *-ase* always indicates an enzyme and the name *lipase* would be used for an enzyme catalysing a reaction which involves a lipid.

An enzyme acts by forming a complex with its substrate (the substance whose reaction it catalyses). The substrate reacts to form products which are then released from the enzyme, leaving it free to catalyse another reaction. Each enzyme has a unique surface structure, a consequence of its globular protein nature, and this provides a precise position (the active site) at which the substrate can join the enzyme molecule to form an enzyme-substrate complex. This intimate contact is maintained until the reaction is complete (see diagram 3.14). The precise and specific fit between enzyme and substrate is sometimes compared with a 'lock-and-key' mechanism.

Enzyme inhibition

Enzymes may be inactivated by substances called inhibitors which interfere

with the catalytic process. This effect may be produced in several different
ways.

(a)   The active site of the enzyme may be blocked by the formation of an
enzyme-inhibitor complex (see diagram 3.14). This is known as competitive
inhibition and occurs when the inhibitor molecule is structurally similar to
the usual substrate of the enzyme.

(b)   The inhibitor may react irreversibly with the enzyme to form an
inactive, non-enzymatic end-product.

(c)   The inhibitor may alter the shape of the enzyme at its point of
activity so that the enzyme-substrate complex cannot form. In some enzymes
this occurs by action at a second site on the enzyme which, although not
enzymatically active, can exert an effect upon the active site. This is known
as an 'allosteric effect' which may be either inhibiting or activating to an
enzyme.

Sulphanilamide is an example of an enzyme inhibitor. It blocks the enzyme
activity necessary for the growth of bacteria and this property accounts for
its efficacy in the treatment of certain bacterially-induced illnesses. Carbon
monoxide inhibits the activity of the enzyme cytochrome oxidase which
plays an essential part in cellular respiration at the end of the respiratory
chain (see page 74). This effect may be lethal.
(Note   Carbon monoxide also combines preferentially with mammalian
haemoglobin, disrupting oxygen transport and producing equally serious
results.)

Factors affecting enzyme activity

Enzyme activity is affected by changes in temperature, in acidity and in
substrate concentration.

Within certain limits an increase in temperature causes an increase in
enzyme activity. Below 37°C, enzymes are progressively inactivated (this
factor accounts for the torpor of cold-blooded animals at low temperatures).
Above this point the rate of enzyme activity increases until an optimum of
about 45°C is reached (depending on the enzyme) after which the rate
rapidly decreases. This diminution of activity results from the irreversible
denaturation of the enzyme protein at high temperatures.

All enzymes have a specific pH point, their optimum pH, at which they
work most effectively. (The term pH refers to hydrogen ion concentration
and measures the acidity or alkalinity of a solution — pH 1 is very acid, pH 7

3.14 *Enzymes*

Enzyme action

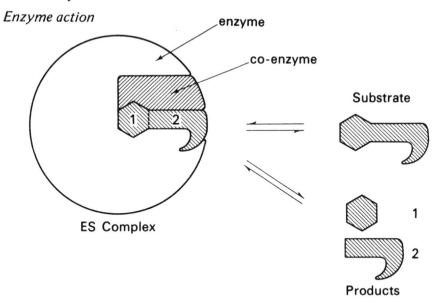

Enzyme + Substrate ⇌ ES Complex ⇌ Enzyme + Products

Enzyme inhibition

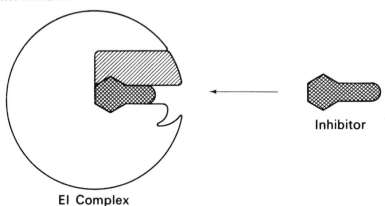

The enzyme inhibitor
complex formed is inactive,
therefore the enzyme
becomes blocked
by the inhibitor

3.15 *Enzyme activity factors*

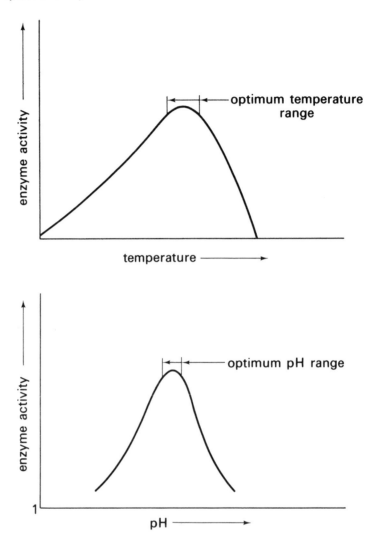

is neutral, pH 14 is very alkaline.) This variation is due to changes in the molecular shape of the enzyme which occur when the pH alters. Most intracellular enzymes work best at approximately pH 7. Extracellular digestive enzymes, which function in the diverse and unusual conditions of the alimentary tract, may be most efficient at more extreme pH values. The enzyme pepsin, operating among the acid stomach contents, has an optimum pH of 2–3.

Enzyme activity is also influenced by the concentration of the substrate. Increasing quantities of substrate will cause an increase in the rate of enzyme activity until a large enough concentration is present to make full use of all the enzymic active sites. Above that point increasing substrate concentration will have no further effect upon enzyme activity.

Enzyme activators — co-factors

In some instances, as already noted, an enzyme must be associated with a smaller molecule in order to function properly. Such a co-factor may be inorganic (e.g. a metal ion) or it may be a small organic co-enzyme molecule (e.g. a nucleotide or a vitamin).

FURTHER READING

General Texts
Clowes, R., *The Structure of Life*, 1967, Penguin
Coult, D. A., *Molecules and Cells*, 1966, Longmans
Rose, S., *The Chemistry of Life*, 1970, Penguin

Scientific American Offprints
Proteins and enzymes
Kendrew, J. C., The 3-D Structure of a Protein, Dec. 1961
Kendrew, J. C., Myoglobin and the Structure of Protein, Dec. 1963
Perutz, M. C., The Haemoglobin Molecule, Nov. 1964
Phillips, D. C., The 3-D Structure of an Enzyme, Nov. 1966

Nucleic acids
Crick, F. H. C., Nucleic Acids, Sept. 1957
Hanawalt, P. C. and Haynes, R. H., Repair of DNA, Feb. 1967
Holley, R. W., Nucleotide Sequence of a Nucleic Acid, Feb. 1966
Mirsky, A. E., The Discovery of DNA, Jul. 1968

More Advanced Texts
Ambrose, E. J. and Easty, D. M., *Cell Biology*, 1970, Nelson
Conn, E. E. and Stumpf, P. K., *Outlines of Biochemistry*, 1967, Wiley
Loewy, A. G. and Siekevitz, P., *Cell Structure and Function*, 2nd edition, 1969, Holt, Rinehart and Winston
Mahler, H. R. and Cordes, E. M., *Biological Chemistry*, 1968, Harper and Row
Watson, J. D., *Molecular Biology of the Gene*, 1965, Benjamin

General Reading
Watson, James D., *The Double Helix*, Weidenfeld and Nicolson; 1970, Penguin

4 Energy Conversion in the Cell

Cells need energy for all the active processes occurring within them. This energy comes primarily from the light energy of the sun which is captured by the chloroplasts of green plant cells and used to build up complex organic molecules from simple inorganic ones. The chemical energy trapped in the complex structure of organic molecules is released when they are broken down during cellular respiration and much of this energy is used by the cell for various types of work.

There are, basically, two categories of cells. One of these contains the self-supporting *autotrophic* cells which can build up their own complex food materials by the process of photosynthesis and then utilise the energy contained in these organic molecules for their cellular activities. The other division includes the dependent *heterotrophic* cells which cannot photosynthesise and so must rely upon obtaining organic food substances from the autotrophic cells (see diagram 4.1).

The energy currency of the cell

Energy in the form of heat is of little use to the cell since it cannot be stored or transported. To be useful to the cell, energy must be packaged in the convenient form of a 'high-energy' chemical bond, a type of cellular energy currency that is easily stored and readily available. The most common energy carrier utilised by the cell is the molecule adenosine diphosphate, ADP. This nucleotide is made up of the base adenine, the sugar ribose and two phosphate groups. When free phosphate and energy are available ADP can link on a further phosphate group by a 'high-energy' bond, which involves an unexpectedly large amount of energy, to form a molecule of adenosine triphosphate, ATP (see page 52 and diagram 4.1).

When energy is made available to the cell much of it is used to build up stocks of ATP from ADP and phosphate. Energy can be stored in this way until it is required to do work. When energy is needed within the cell, ATP is degraded to ADP and phosphate with the liberation of chemical bond energy which is immediately utilised.

4.1 Autotrophs and heterotrophs

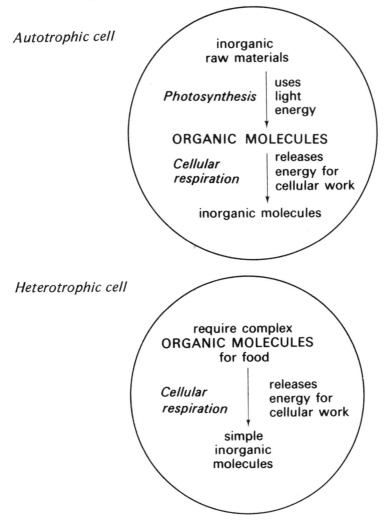

Autotrophic cell

inorganic
raw materials

Photosynthesis | uses
light
energy

ORGANIC MOLECULES

*Cellular
respiration* | releases
energy for
cellular work

inorganic molecules

Heterotrophic cell

require complex
ORGANIC MOLECULES
for food

*Cellular
respiration* | releases
energy for
cellular work

simple
inorganic
molecules

Energy currency of the cell

$$A{-}P{\sim}P{\sim}P \rightleftharpoons A{-}P{\sim}P \;+\; P \;+30000 \text{ joules/mole}$$

ATP   ADP   phosphate   energy

~ = a high energy bond

Energy conversion

Cells must do work in order to maintain their integrity and perform their specialised tasks. The basic machinery of cellular function is concerned with the supply of energy, in the convenient form of ATP, for use in these tasks. This energy supply is ultimately derived from solar energy which is first converted by autotrophic cells into energy-rich organic molecules using the unique process of photosynthesis. Once stored in this form the energy is available for cellular purposes as all cells can unlock the storehouse and repackage the energy into useful ATP molecules by means of the ubiquitous, intricate catabolic operation of cellular respiration.

All cells are necessarily energy converters, but, although some bacteria obtain energy from environmental chemicals by chemosynthesis, only the specially-endowed photosynthetic cells are capable of trapping solar energy and utilising it for their own designs (see diagram 4.2).

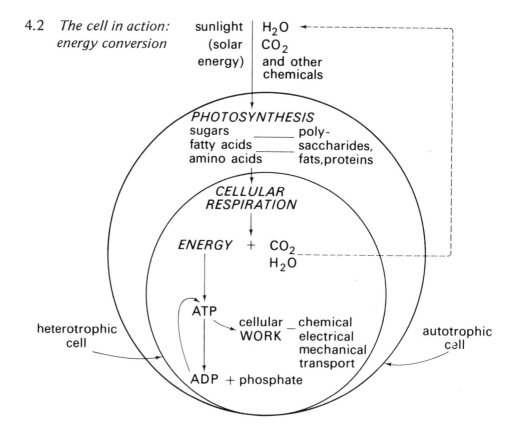

4.2   *The cell in action: energy conversion*

sunlight (solar energy) | H_2O
CO_2
and other chemicals

PHOTOSYNTHESIS
sugars ———— poly-
fatty acids ———— saccharides,
amino acids   fats, proteins

CELLULAR RESPIRATION

ENERGY + CO_2
H_2O

ATP

cellular WORK — chemical
electrical
mechanical
transport

ADP + phosphate

heterotrophic cell

autotrophic cell

Photosynthesis

The process of photosynthesis is fundamental to all living things as it is the basic step by which organic molecules are manufactured from inorganic materials. Only plant cells containing chlorophyll are able to photosynthesise.

Photosynthesis is usually represented by this simple equation:

$$CO_2 + H_2O + \text{sunlight} \longrightarrow \underset{\text{carbohydrate}}{(CH_2O)} + O_2$$

However this equation describes only the total input and output of the process and fails to indicate the complex series of intermediate reactions which are involved.

In order to discuss the complicated process of photosynthesis it is helpful to divide it into three consecutive major stages (see diagram 4.3):

1 Photolysis — the light reaction which splits water molecules into their component atoms and stores some light energy in the form of ATP.

2 Carbon dioxide fixation — a dark reaction in which carbon dioxide is combined with the hydrogen released by photolysis to form a simple triose carbohydrate.

3 Carbohydrate conversion — in which the simple carbohydrate is changed into hexose monosaccharides and the polymerised carbohydrates.

4.3   *Photosynthesis*

a Basic input/output reaction

$$CO_2 + H_2O + \text{Sunlight} \longrightarrow \underset{\text{carbohydrate}}{(CH_2O)} + O_2$$

b Backbone reactions

1 Photolysis

sunlight
↓
$$\underset{\text{chlorophyll}}{\text{excited}} + 2H_2O \longrightarrow \underset{\text{chlorophyll}}{\text{normal}} + (2H_2) + O_2 + \underset{\text{energy}}{\text{available}}$$

2 Carbon dioxide fixation

$$(2H_2) + CO_2 + \text{energy} \longrightarrow \underset{\text{carbohydrate}}{(CH_2O)} + H_2O$$

3 Carbohydrate conversion

$$(CH_2O) \longrightarrow \text{more complex carbohydrates, fats and proteins.}$$

1 Photolysis or the Hill Reaction

The first stage of photosynthesis uses light energy to split water molecules into their component hydrogen and oxygen atoms. The hydrogen atoms are incorporated into the carbohydrate formed in the later part of the process. The oxygen atoms are not used and are given off as a by-product (see diagram 4.3).

The light energy for photolysis is trapped by the green pigment chlorophyll which is found in the grana of the chloroplasts of green cells (see page 22). Chlorophyll molecules are sensitive to light and can absorb light energy, causing certain electrons to become so 'excited' that they leave the molecule. These energy-rich 'excited' electrons are involved in the reactions which split the water molecules. The released hydrogen atoms are transferred to a 'hydrogen acceptor' molecule called NADP (nicotinamide adenine dinucleotide phosphate) forming reduced NADP or $NADPH_2$ and some of the liberated energy is stored in ATP molecules built from ADP and phosphate. The oxygen atoms released from the water molecules play no further part in photosynthesis and are given off as a waste product.

While this account of photolysis is essentially true, it is also very simplified. Readers with a chemical background will be aware that reduction reactions, although they often involve the removal of oxygen or the addition of hydrogen, may also occur solely by the addition of electrons. Biological oxidation/reduction reactions are frequently accomplished by means of the transfer of electrons from one substance to another. Hydrogen carriers, such as NADP, which exist in oxidised and reduced forms may actually carry not hydrogen but electrons and should more accurately be called 'electron carriers'.

The light reactions of photosynthesis are more complex than they initially appear. Two types of chlorophyll, *a* and *b*, are found in the chloroplasts of green plants where they are thought to perform rather different but co-ordinated jobs. Light-excited, energy-rich electrons from chlorophyll *b*, derived from the hydroxyl ions of water, pass along a complex chain of electron transporters (probably similar to those found in mitochondria, see page 72) where their energy is tapped off to produce ATP from ADP and phosphate. These low-energy electrons are then picked up by chlorophyll *a* to become excited by light once more and to pass via an electron carrier called ferrodoxin to NADP which is reduced to $NADPH_2$. Energy-rich electrons from chlorophyll *b* therefore generate ATP while those from chlorophyll *a* give rise to reduced NADP — both these products are used in the later stages of photosynthesis to reduce CO_2 to carbohydrate (see diagram 4.4).

4.4   *Photosynthesis — detailed reactions*
1 Photolysis

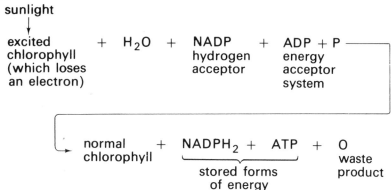

Photolysis — a current view

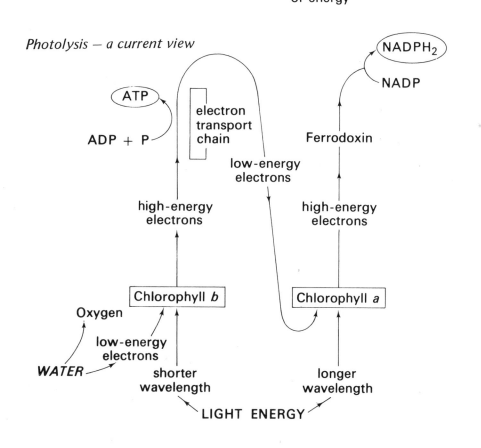

2 Carbon dioxide fixation or the Calvin cycle

The useful products of photolysis, the reduced hydrogen carrier NADPH$_2$ and the stored energy of the ATP molecules, are used in the second part of photosynthesis. During this stage hydrogen is combined with carbon dioxide to form a simple carbohydrate in a series of reactions which make up one of the many cyclical processes occurring in the cell.

Essentially carbon dioxide enters the cycle at the point where it may combine with a 'carbon dioxide acceptor' molecule. Then hydrogen from the hydrogen carrier joins the cycle and, after a complex series of molecular re-arrangements, carbohydrate molecules leave the cycle and the 'carbon dioxide acceptor' is reformed. This complicated cycle actually involves three complete turns before one carbohydrate molecule is formed and at least ten major steps occur during the process (see diagram 4.5).

The constantly recycling 'carbon dioxide acceptor' molecule is ribulose diphosphate, RDP, a phosphorylated ketopentose. When this molecule combines with carbon dioxide two molecules of phosphoglyceric acid, PGA, are formed which, by a series of ATP-requiring reactions and the addition of hydrogen from the carrier NADPH$_2$, are transformed into a triose phosphate, phosphoglyceraldehyde, PGAL, (for formulae see diagram 4.5). This molecule is a simple triose monosaccharide with the general formula $C_3H_5O_3P$. Only one-sixth of the PGAL formed can be used as a carbohydrate source as the remaining five-sixths must pass on round the cycle through a complex series of products until 5—carbon RDP is finally reformed, ready once again to act as a 'carbon dioxide acceptor' and to maintain the cycle.

3 Carbohydrate conversion

Phosphoglyceraldehyde, PGAL, does not accumulate in the grana of the chloroplasts since it is rapidly used up by the cell. It may act as a respiratory substrate to provide energy for its own cell or it may be converted into another carbohydrate product (see diagram 4.6). Alternatively it may be transported outside the cell to other parts of the organism, in which case it will be converted to a less reactive substance such as sucrose, fructose or glucose. If it is to be stored for future use it will be changed into the polysaccharides starch or inulin.

Chloroplasts and photosynthesis

Photosynthesis occurs in the chlorophyll-containing chloroplasts of green

4.5   *Photosynthesis — detailed reactions*

2 Carbon dioxide fixation

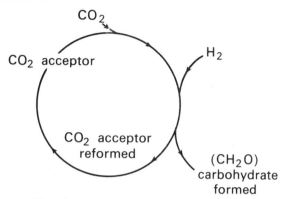

The Basis of the Calvin Cycle

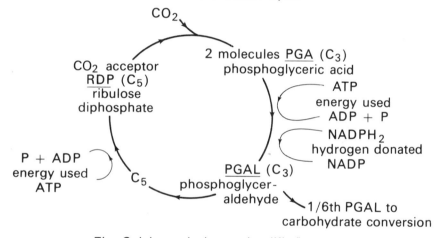

The Calvin cycle (*very simplified*)

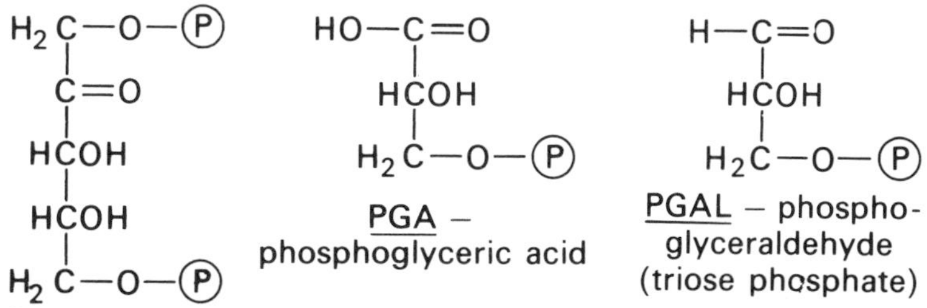

H$_2$C—O—Ⓟ   HO—C=O   H—C=O
   | | |
  C=O   HCOH   HCOH
   | | |
 HCOH   H$_2$C—O—Ⓟ   H$_2$C—O—Ⓟ
   |
 HCOH   PGA —   PGAL — phospho-
   | phosphoglyceric acid   glyceraldehyde
H$_2$C—O—Ⓟ   (triose phosphate)

RDP — ribulose diphosphate
     Participants in the Calvin cycle — RDP, PGA and PGAL

4.6   *Photosynthesis – detailed reactions*

3 Pathway of carbohydrate conversion

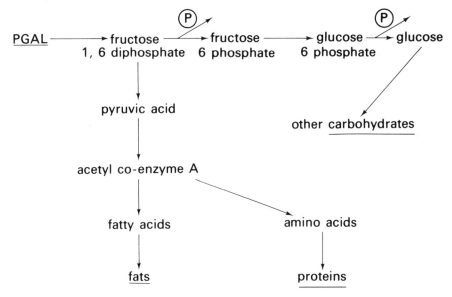

plant cells (for chloroplast structure see page 22). The molecules associated with the light reactions of photosynthesis occur in ordered array tightly bound to the membranes of the grana. These molecules include chlorophyll and the enzymes involved in ATP synthesis, NADP reduction and accessory reactions. The enzymes utilised in the dark reactions of carbon dioxide fixation and in the synthesis of carbohydrates are found in the stroma of the chloroplast.

Cellular respiration

During cellular respiration complex organic molecules are broken down to inorganic components so that the chemical energy locked in their molecular structure is liberated in small packets which may be utilised by the cell. Since all cells need energy, this is a universal process. Respiration is an oxidative (aerobic) mechanism in most cells but in some, particularly those of bacteria and fungi, it may take place anaerobically.

Respiratory breakdown of carbohydrates

Glucose is the most widely used respiratory substrate and it will be taken as the starting point for the process. Other carbohydrates as well as fats and proteins may be broken down to release energy and their role will be discussed later (see page 75).

The energy built into the molecular structure of glucose is extracted during cellular respiration by dismantling the molecule, step by step, and conserving much of the released energy in the phosphate bond energy of ATP. This intricate process is remarkably efficient. During aerobic respiration nearly 50 per cent of the available energy of the glucose molecule is recovered as utilisable phosphate bond energy while the remainder is dissipated as heat.

Cellular respiration takes place in two major phases:

1 Glycolysis — the preparatory phase which takes place in the cytoplasm and splits the 6—carbon glucose molecule into two 3—carbon molecules.

2 The Krebs cycle — this cyclical process produces carbon dioxide, water and energy from the 3—carbon molecule made in glycolysis. The cycle takes place in the mitochondria and requires the presence of oxygen.

1 Glycolysis

The C_6 glucose molecule is split into two C_3 pyruvic acid molecules during glycolysis. This preparatory phase of respiration occurs in the cellular cytoplasm and does not require oxygen (see diagram 4.7).

Under anaerobic conditions pyruvic acid is converted to lactic acid in animal cells and to ethyl alcohol and carbon dioxide in plant cells. At this point the process stops, having released only about 3 per cent of the stored energy of the glucose molecule which has been used to generate two molecules of ATP. Anaerobic cells must function on this small energy yield. When aerobic conditions prevail the pyruvic acid produced by glycolysis is broken down to C_1 molecules of carbon dioxide in the Krebs cycle, yielding a great deal more energy.

Glycolysis is a complex process. It involves at least eleven reactions, each catalysed by a different enzyme, in order that the energy of the glucose molecule may be efficiently extracted in convenient packages. The initial stage activates the glucose molecule by phosphorylation (see page 39), an energy-requiring step which forms glucose-6-phosphate. The activated glucose molecule then participates in a series of reactions which eventually produces two molecules of pyruvic acid (for details of these reactions see diagram 4.8). The energy released by the process is stored in two molecules of ATP and in two molecules of an energy-rich compound $NADH_2$ (NAD, nicotinamide adenine dinucleotide, is a hydrogen carrier, see page 52).

4.7   *Respiration*

Basic input/output equations

1 Aerobic respiration

$$C_6H_{12}O_6 + 6\ O_2 \longrightarrow 6\ CO_2 + 6\ H_2O + \text{ energy}$$

2 Anaerobic respiration

$$C_6H_{12}O_6 \longrightarrow 2\ CH_3CHOH.COOH + \underset{\text{lactic acid}}{} \text{ energy}$$

lactic acid   in <u>animal</u>
                                       cells

$$C_6H_{12}O_6 \longrightarrow 2\ C_2H_5OH + 2\ CO_2 + \text{ energy}$$

ethyl alcohol   in <u>plant</u>
                                       cells

Backbone diagram

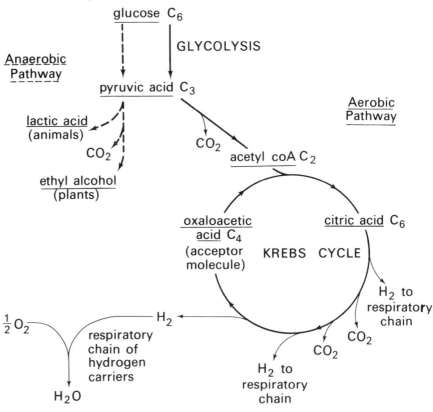

During anaerobic respiration the two molecules of $NADH_2$ reduce pyruvic acid to lactic acid or ethyl alcohol. In aerobic respiration they are later oxidised to NAD by means of the respiratory chain (see next section) with the formation of three molecules of ATP for each molecule of $NADH_2$ oxidised. For this reason the energy yield from glycolysis is higher in the presence of oxygen. One glucose molecule undergoing glycolysis will yield the energy to generate two molecules of ATP under anaerobic conditions but this score is increased to eight molecules of ATP when oxygen is available.

4.8   Respiration details

1 Glycolysis (Embden Meyerhof pathway)

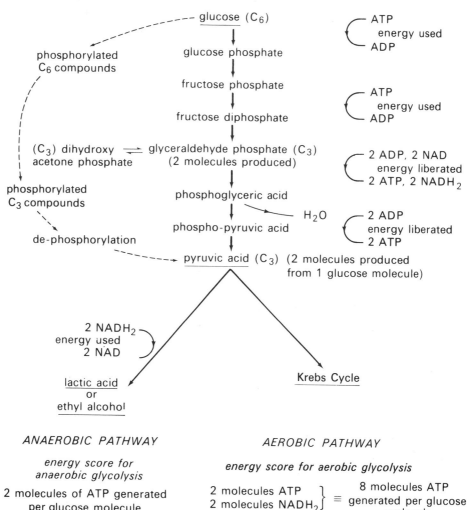

ANAEROBIC PATHWAY

energy score for
anaerobic glycolysis

2 molecules of ATP generated
per glucose molecule

AEROBIC PATHWAY

energy score for aerobic glycolysis

2 molecules ATP
2 molecules $NADH_2$ $\Bigg\}$ $\equiv$ 8 molecules ATP generated per glucose molecule

2 The Krebs cycle (tricarboxylic acid cycle or citric acid cycle)

Once glucose has been broken down to pyruvic acid by glycolysis, aerobic cells can use the Krebs cycle to extract a large proportion of the energy that remains in the molecule. Before this mitochondrial cycle can begin, however, pyruvic acid must be decarboxylated (by removal of a carboxyl group) in a reaction with co-enzyme A, which acts as both an acetyl carrier and an energy carrier. The acetyl group from the C_3 pyruvic acid molecule is attached to co-enzyme A to form C_2 acetyl co-enzyme A, carbon dioxide is given off and hydrogen passes to the carrier NAD ($NADH_2$ will be later oxidised to release energy).

$$CH_3CO.COOH + CoASH + NAD \longrightarrow CH_3CO.SCoA + CO_2 + NADH_2$$
$$\text{pyruvic acid} \quad \text{co-enzyme A} \qquad\qquad \text{acetyl} \quad \text{co-A}$$

C_2 acetyl co-enzyme A is the starting point of the Krebs cycle. It combines with its acceptor molecule, C_4 oxaloacetic acid, to form C_6 citric acid which is then broken down in a complex series of reactions. Citric acid is degraded first to C_5 ketoglutaric acid and then to C_4 succinic acid. Eventually the C_4 molecule is converted back to the acceptor molecule, oxaloacetic acid, so that the cycle may turn once again (for details of these reactions see diagram 4.9).

In the course of the Krebs cycle, carbon dioxide is given off and high-energy hydrogen ions are released which are picked up by the hydrogen carriers NAD or FP. The reduced carriers are subsequently oxidised in the respiratory chain where energy is transferred to the bonds of ATP molecules. The carriers of the respiratory chain always work in a specific order. The sequence involves first NAD, then the flavoproteins, FP (which may be either flavin mononucleotide, FMN, or flavin adenine dinucleotide, FAD), next the quinones whose action is poorly understood and, finally, a series of cytochromes (iron-containing chromoproteins) which pass the hydrogen ions* to molecular oxygen with the formation of water (see diagram 4.10). It is only at this point in the aerobic respiratory process that oxygen is required but if it is not available the Krebs cycle ceases to turn and C_3 pyruvic acid, not C_1 carbon dioxide, is the end-product.

*Note.
As noted in photosynthesis, reduction may involve not the addition of hydrogen ions but the addition of electrons. At some stages the respiratory chain acts as a hydrogen carrier but at many points it transports only electrons.

4.9   Respiration details

2 Krebs cycle

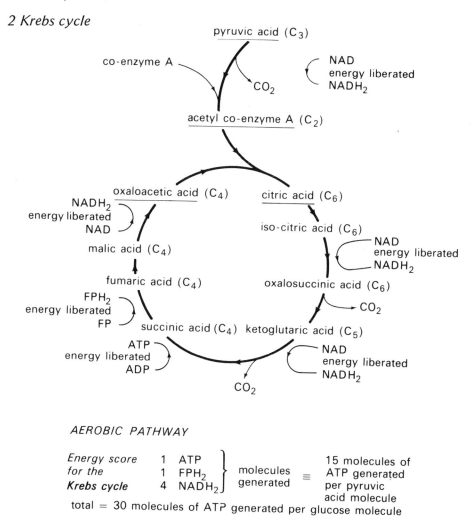

AEROBIC PATHWAY

Energy score   1   ATP   ⎫
for the   1 FPH_2 ⎬ molecules $\equiv$ 15 molecules of
Krebs cycle 4 $NADH_2$ ⎭ generated   ATP generated
               per pyruvic
               acid molecule
   total = 30 molecules of ATP generated per glucose molecule

The energy yield from respiration

The passage of a hydrogen ion down the respiratory chain yields enough energy to synthesise three molecules of ATP if it starts at $NADH_2$; this is reduced to two molecules of ATP if the ion starts at FPH_2 (see diagram 4.10). It is therefore possible to calculate the total energy yield for aerobic and anaerobic respiration in terms of the number of ATP molecules generated for each glucose molecule respired.

		Energy yield per Glucose molecule respired
Anaerobic respiration		
Glycolysis		2 molecules ATP
Aerobic respiration		
Glycolysis		2 molecules ATP
	2 molecules $NADH_2$ =	6 molecules ATP
Krebs cycle (1 glucose molecule produces 2 pyruvic acid molecules)		2 molecules ATP
	2 molecules FPH_2 =	4 molecules ATP
	8 molecules $NADH_2$ =	24 molecules ATP
		38 molecules ATP

4.10 *Respiration details*

3 Respiratory chain of hydrogen carriers

The sequence of carriers

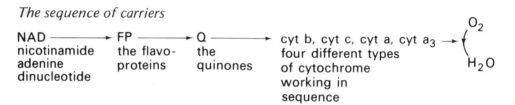

NAD
nicotinamide adenine dinucleotide
FP
the flavo-proteins
Q
the quinones
cyt b, cyt c, cyt a, cyt a3
four different types of cytochrome working in sequence
O2
H2O

The passage of hydrogen down the respiratory chain

H2
NAD
NADH2
FPH2
FP
Q
QH2
reduced
cytochrome
oxidised
½O2
H2O
energy liberated
ATP formed
energy liberated
ATP formed
energy liberated
ATP formed

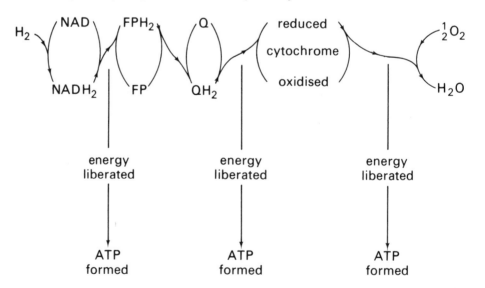

Mitochondria and cellular respiration

Mitochondria are known as the 'power houses' of the cell because a large part of aerobic cellular respiration takes place inside them. The preparatory glycolytic phase occurs in the cytoplasm surrounding the mitochondria. Pyruvic acid then enters the mitochondria to be degraded to carbon dioxide and water. The enzymes associated with the Krebs cycle are found in solution in the internal matrix of the mitochondrion and the components involved in the respiratory chain are firmly bound in ordered sequence to the lipoprotein membrane lining the inner surface of the cristae (for mitochondrial structure see page 19).

Pentose shunt

Alternative pathways exist for the oxidation of glucose and the most important of these is the 'pentose shunt' which bypasses glycolysis and the Krebs cycle. This pathway travels via a 5-carbon sugar, ribulose-5-phosphate, and it is the major source of ribose and deoxyribose sugars, components of the nucleic acids and important nucleotides such as ATP, NAD and FAD. Ribulose-5-phosphate is the starting point for the synthesis of many other sugars and its phosphorylated derivative, ribulose diphosphate, acts as the carbon dioxide acceptor molecule in photosynthesis (see page 66).

Fuels for cellular respiration

Under normal conditions most cells supply their energy requirements by the oxidation of sugars but fats and proteins also serve as energy sources.

1 Carbohydrates

Most cells receive a supply of glucose which can be oxidised directly by the pathways already described. Stored carbohydrates in the form of poly-saccharides must be hydrolysed to their monosaccharide components before they can be respired. Animal cells containing stored glycogen, such as liver and muscle cells, possess enzymes which unlink the glucose units of the glycogen chain and convert them to glucose-6-phosphate, a participant in the respiratory process (see diagram 4.11).

2 Fats

In general, fats are used as a major energy source only when glucose is in short supply. Such a situation may be due to momentary or prolonged starvation, disease (e.g. diabetes) or other unusual conditions.

Fats for respiration are first hydrolysed to fatty acids and glycerol. The glycerol is then converted to dihydroxyacetone phosphate which feeds into the glycolysis pathway at an intermediate point, glyceraldehyde-3-phosphate. Fatty acids are broken into C_2 acetate segments which combine with co-enzyme A to form acetyl co-enzyme A and then can be further oxidised in the Krebs cycle (see diagram 4.11).

4.11   *Respiration details*

4 Fuels for cellular respiration

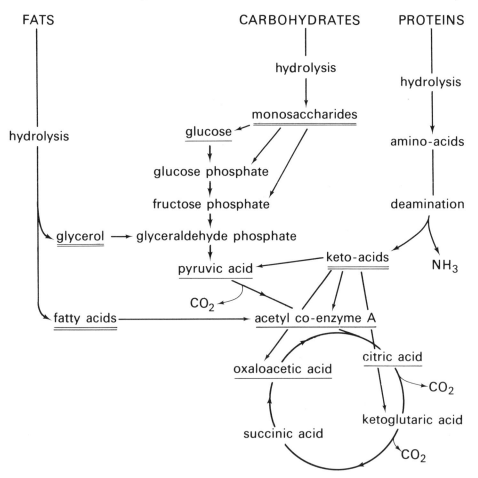

3 Proteins

Proteins are so vital to the structure and function of the cell that they are used as a major source of energy only as a last resort when fat stocks have run low, except in some animals with a very high protein diet.

Proteins are hydrolysed to their constituent amino acids which are then deaminated to yield keto-acids and ammonia. The ammonia is excreted from the cell as free ammonia, uric acid or urea. The keto-acids may be of three types — pyruvic acid, ketoglutaric acid or oxaloacetic acid — all of which are participants in the respiratory pathways, or they may resemble fatty acids and feed into the Krebs cycle at acetyl co-enzyme A (see diagram 4.11).

FURTHER READING

General Texts
Baron, W. M. M., *Organization in Plants*, 1967, Arnold
Goldsby, R. A., *Cells and Energy*, 1967, Macmillan
Loewy, A. G. and Siekevitz, P., *Cell Structure and Function*, 2nd edition, 1969, Holt Rinehart, and Winston
McElroy, W. D., *Cell Physiology and Biochemistry*, 2nd edition, 1964, Prentice Hall
Rose, S., *The Chemistry of Life*, 1970, Penguin
Scientific American Offprints
Arnon, D. I., The Role of Light in Photosynthesis, Nov. 1960
Bassham, J. A., The Path of Carbon in Photosynthesis, Jun. 1962
Levine, R. P., The Mechanism of Photosynthesis, Dec. 1969
Rabinowitch, E. I. and Gorindjee, The Role of Chlorophyll in Photosynthesis, July 1965
Lehninger, A. L., How Cells Transform Energy, Sept. 1961

5 The Cell at Work: 1

Cells are basically instruments of energy transformation. The process of respiration makes energy available to the cell which may be used to do work of many kinds. In order to maintain itself, the cell must continually use energy to repair and replace its own structure by the demolition of worn parts and their reconstruction with materials synthesised in the cell. Even more energy will be used in cell growth and division. Substances must be moved both within and between cells, and specialised activities, such as muscle contraction, nervous conduction, sensory perception and light emission, may take place. All these operations involve the utilisation of energy stored in the high energy bonds of ATP.

Controlling the internal environment of the cell

Selective barriers, the cell membranes, regulate the internal environment of the cell and maintain its integrity while controlling the traffic of materials between the cell and its external surroundings. The cell will require food, inorganic compounds, oxygen and other essential substances, and it must be able to expel secretory and waste products.

Different solutions cross the membrane barriers with varying facility. In general, small molecules penetrate the cell with greater ease than large ones, although this is not always the case (see page 79). Large organic molecules must be hydrolysed to smaller constituents before they can enter the cell (this is the fundamental function of the digestive system of higher animals).

Materials can enter and leave the cell by one or more of four main methods:

1 diffusion
2 osmosis
3 active transport
4 pinocytosis and phagocytosis.

1 Diffusion

Diffusion is the process whereby a dissolved substance will gradually spread through a solution until its molecules are evenly distributed throughout that solution.

    If the cell membrane is fully permeable to a certain substance its molecules will tend to move across the membrane regions of high concentration to those of low concentration. Many inorganic and organic materials move across membranes by diffusion. In some instances the passive forces of diffusion across a membrane are assisted by enzymes present in that membrane. The entry of glucose into the human erythrocyte is an example of this 'enzyme-controlled permeability'.

2 Osmosis

The cell membrane is not permeable to all substances. When the membrane acts as a selective barrier, allowing the passage of some molecules but restricting others, then an osmotic system may be set up. In such a system two aqueous solutions of unequal concentrations are separated by a membrane permeable only to the solvent (water). If solvent molecules only can move across the membrane then these will pass from the more dilute to the more concentrated solution in an attempt to equalise the concentration difference and restore equilibrium. The pressure which may be exerted by this movement is called the 'osmotic pressure'. Osmosis will occur whenever molecules cannot cross the cell membrane, causing the solvent water to migrate through the membrane. This is one agency by which water is distributed from cell to cell and it is of particular importance in plant tissues.

3 Active transport

Cell membranes are selectively permeable, being most permeable to water and dissolved gases while other molecules and ions penetrate more slowly (most membranes are less permeable to anions than cations). A very rapid exchange of important substances between the external and internal environments is often necessary for the efficient functioning of the cell so a specialised form of transport has developed to carry this out. The process may occur in either direction across a membrane; if it reinforces diffusion it is called 'enzyme controlled permeability' but if materials are carried against the tides of diffusion then work must be done, energy must be consumed and the operation is known as 'active transport'.

Some marine algae can actively accumulate iodine until its cellular concentration is more than one million times greater than that of the surrounding seawater. Many other cells, such as the human erythrocyte, tend to concentrate potassium ions and to eliminate sodium ions against the concentration gradient. This active mechanism, the 'sodium pump', is responsible for the extrusion of sodium ions from the cell and probably also for the cellular concentration of potassium ions. It is suggested that this linked $Na+/K+$ transport involves a 'carrier' molecule lying in the cell membrane which combines with the ions and ferries them across the membrane barrier. The carrier molecules are thought to be part of the protein fraction of the lipoprotein plasma membrane. Like all enzyme activity, protein 'carrier' action is very specific.

Studies on the bacterium *Escherichia coli* have demonstrated a protein lactose-carrier molecule in the plasma membrane which enables the bacterium to develop an internal lactose concentration 500 times greater than that of the external surroundings. The carrier is thought to act like a revolving door which allows the entry of molecules only with a specific size and shape, in this case lactose. The passenger molecule then swings with the 'carrier' door to the inside surface of the membrane and so to the cell's interior. Metabolic energy is needed to revolve the 'carrier' door back to the open position on the outside of the cell (see diagram 5.1).

4 Pinocytosis and phagocytosis

White blood corpuscles and other amoeboid cells can take substances in bulk into the cell by the formation of membrane invaginations which pinch off and move into the cytoplasm. When fluid engulfment occurs the process is called pinocytosis while phagocytosis is the inclusion of larger, solid particles by means of pseudopodia from the cell surface.

Vesicles formed by these processes, either pinocytic vesicles or phagosomes, fuse with one or more lysosomes (see page 23) to make digestive vacuoles. Here the ingested material is hydrolysed by the enzymes of the lysosome into small molecules and the useful products pass into the cytoplasm. Any undigested matter forms a residual body which may be deposited in the cytoplasm or expelled from the cell.

5.1   *The carrier concept in active transport*

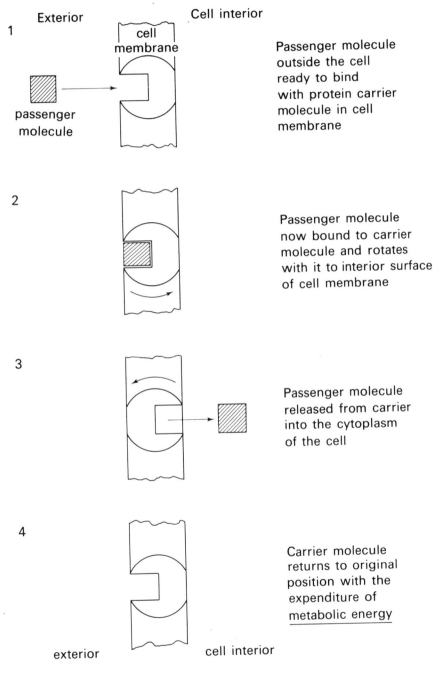

1   Passenger molecule outside the cell ready to bind with protein carrier molecule in cell membrane

2   Passenger molecule now bound to carrier molecule and rotates with it to interior surface of cell membrane

3   Passenger molecule released from carrier into the cytoplasm of the cell

4   Carrier molecule returns to original position with the expenditure of metabolic energy

Biosynthesis

Biosynthesis, the production of molecules by living matter, is a vital cellular activity and energy must be used to maintain the constant supply of nucleic acids, proteins, carbohydrates and lipids required in the structure and functions of growing and mature cells.

Protein synthesis

Proteins play two important roles in cells; they supply the building units for many cellular components and they act as enzymes, catalysing myriads of biochemical reactions. Since the proteins of one cell may be very different from those essential to another, each cell must be able to synthesise its own specific proteins from appropriate amino-acid sub-units. Instructions for specific protein synthesis in each cell are stored in the DNA code of the chromosome strands and confined within the nuclear membrane to protect them from destruction by cytoplasmic enzymes. But proteins are actually assembled on the ribosomes of the cytoplasm, so a special messenger molecule called 'messenger RNA' carries a transcript of the DNA instructions from the nucleus to the ribosomal workshops where they are translated into terms of the sequences of amino-acids within protein macromolecules.

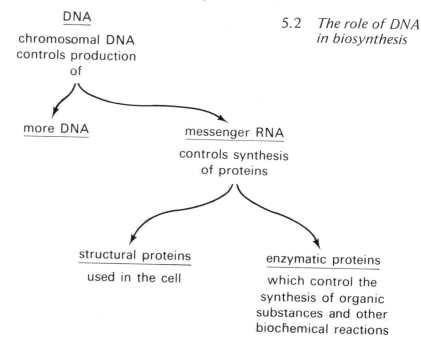

DNA

chromosomal DNA
controls production
of

more DNA

messenger RNA

controls synthesis
of proteins

structural proteins

used in the cell

enzymatic proteins

which control the
synthesis of organic
substances and other
biochemical reactions

5.2   *The role of DNA
in biosynthesis*

The genetic code of DNA is embodied in the precise sequence of nitrogen bases along the length of the DNA molecule, which in turn prescribe the sequence in which amino-acids link to build a particular protein chain. The four different nucleotide bases of DNA — adenine, thymine, cytosine and guanine — act as the four 'letters' of a short genetic code alphabet. Code words or 'codons', made from groups of three nucleotide letters, represent the twenty different amino-acids of protein manufacture. In theory, sixty-four triplet words could be constructed using all the possible combinations of the four nucleotide letters. While only twenty of these triplets are needed as codons for the twenty amino-acids, the extensive experimentation which cracked the genetic code has shown that it is somewhat degenerate as one amino-acid may be represented by more than one codon (for details of these fascinating experiments see Further Reading, especially Crick, Nirenberg and Watson).

The nucleus of the cell acts as a library housing 'master copies' of the DNA code instructions for protein synthesis. When a certain protein is needed in the cell the coded genetic instructions relating to that protein are copied from the DNA 'master' on to expandable messenger RNA strands. This is achieved by separation of the two DNA strands, one of them acting as a template for a single strand of messenger RNA so that, by precise point-to-point base pairing, the molecular instructions are transcribed from DNA to

Table 5.1   *Types of nucleic acid found in the cell*

Name	Type of molecule	Location	Function
DNA Deoxyribo-nucleic acid	Macromolecule in shape of double helix, with many thousands of sub-units	Mainly in nucleus, also in mitochondria and chloro-plasts	Acts as a store of coded instructions for synthesis of all proteins required by the cell.
mRNA Messenger ribonucleic acid	Single-stranded polymer, with hundreds of sub-units	In nucleus and cyto-plasm, especially ribosomes	Made on the DNA template, it carries coded instructions for synthesis of one or more proteins from nucleus to ribosomes.
rRNA Ribosomal ribonucleic acid	Molecule very closely bound to protein fraction	Only in ribosomes	Forms part of ribosome structure. Helps in locating mRNA correctly on ribosome surface.
tRNA Transfer ribonucleic acid	Single-stranded polymer of less than one hundred sub-units	In the cytoplasm	Many kinds of tRNA act as amino-acid carriers. Take specific amino-acid from cytoplasm to mRNA template on ribosome.

Table 5.2   *Plan of protein synthesis*

1   Genetic DNA unwinds and unzips to act as template for mRNA formation
2   mRNA leaves nucleus for ribosome or polysome
3   mRNA locates accurately on ribosome surface with help of rRNA
4   tRNA brings required amino-acids to fit the appropriate places on the mRNA template
5   Amino-acids join up by peptide links to form polypeptide chains
6   The protein is built up and leaves the mRNA strand which acts as template for the production of another identical protein molecule

complementary mRNA (see diagram 5.3). The mRNA working transcript of the instructions then leaves the nucleus, probably through the pores of the nuclear membrane (see page 26). It moves into the cytoplasm where its encoded instructions are translated into protein structure on a ribosome. mRNA becomes attached to the surface of one or more ribosomes; several ribosomes may co-operate to produce one protein molecule and are known as a polyribosome or polysome. Another type of RNA (rRNA) is found in ribosomes and seems to ensure that the mRNA strand is correctly aligned over its ribosomes. A third type of RNA also plays a part in protein synthesis. This is a smaller molecule called transfer RNA, tRNA, which acts as linkman between mRNA and amino-acid units. Each tRNA molecule has an anti-codon site which binds to one particular nucleic acid codon. The tRNA molecule also has a binder site for its own particular amino-acid so it can recognise and pick up a specific amino-acid and carry it to an appropriate point on the mRNA template. As polypeptide strands are constructed to the specifications of the mRNA template they are hooked together to form the required protein molecule. At the completion of this complex, energy-consuming process the newly-synthesised protein breaks away from the ribosome surface, leaving the mRNA blueprint free to manufacture another identical molecule. The expendable mRNA strand is usually destroyed by enzymes after several protein molecules have been made.

Plant cells are able to synthesise all the amino-acids that they need from intermediates in the photosynthetic and respiratory pathways and from other precursors. Animal cells cannot synthesise all the amino-acids they require so those they cannot make must be obtained from outside the cell as essential dietary constituents.

5.3 *Coding patterns – DNA to protein*

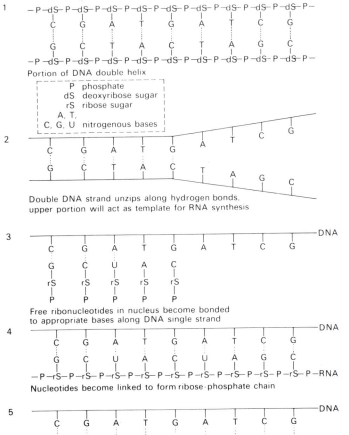

1

–P–dS–P–dS–P–dS–P–dS–P–dS–P–dS–P–dS–P–dS–P–dS–P–
C G A T G A T C G
G C T A C T A G C
–P–dS–P–dS–P–dS–P–dS–P–dS–P–dS–P–dS–P–dS–P–dS–P–

Portion of DNA double helix

P	phosphate
dS	deoxyribose sugar
rS	ribose sugar
A, T, C, G, U	nitrogenous bases

2

C G A T G A T C G
G C T A C T A G C

Double DNA strand unzips along hydrogen bonds, upper portion will act as template for RNA synthesis

3

DNA
C G A T G A T C G
G C U A C
rS rS rS rS rS
P P P P P

Free ribonucleotides in nucleus become bonded to appropriate bases along DNA single strand

4

DNA
C G A T G A T C G
G C U A C U A G C
–P–rS–P–rS–P–rS–P–rS–P–rS–P–rS–P–rS–P–rS–P–rS–P–RNA

Nucleotides become linked to form ribose-phosphate chain

5

DNA
C G A T G A T C G
G C U A C U A G C
mRNA

mRNA strand is produced with its sequence of bases determined by that of DNA

6

mRNA
G C U A C U A G C
tRNA1 tRNA2 tRNA3
aa1 aa2 aa3

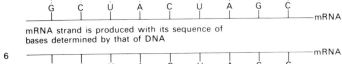

mRNA strand moves to the cytoplasm where it acts as a template for protein synthesis on a ribosomal surface. tRNA acts as an amino-acid carrier.

7

mRNA
G C U A C U A G C
aa1 aa2 aa3 polypeptide

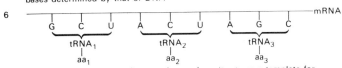

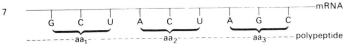

Amino-acid units polymerise to form polypeptide chains. The amino-acid sequence of these chains is determined by mRNA and therefore it is ultimately determined by DNA.

Nucleic acid synthesis

Genetic DNA acts as a template for its own synthesis and so ensures that the vital information encoded in its molecule passes intact from one generation of cells to another.

The mechanism of this process is poorly understood but it is known to involve the untwisting of the DNA double helix and its unzipping into two separate strands. Each of the single strands then links free nucleotides to itself by means of hydrogen bonds and so builds up a new complemental strand (see diagram 5.4).

DNA replication is usually very accurate but changes in DNA structure sometimes do occur. These alterations are known as gene mutations and result from the imperfect copying which may occur during the formation of new DNA. (Chromosomal mutations are changes in the total amount or position of genetic material.)

Several types of gene mutation may be identified. A nucleotide may be chemically changed so that it can establish abnormal base-pair relationships during DNA replication and so change a triplet codon. This may result from molecular rearrangement (tautomerism) of the base, as in the case of the adenine molecule where an unusual, tautomeric form of the molecule will bond with cytosine instead of making the usual adenine/thymine pair bond. On the other hand, the whole nucleotide sequence of a DNA chain may be altered because nucleotides, or blocks of nucleotides, are lost or become erroneously inserted into a nucleotide strand (see diagram 5.5). In this way the codon sequence of a DNA strand may become considerably changed.

Mutations occur spontaneously, possibly as an outcome of normal environmental factors such as cosmic radiation, but the mutation rate can be greatly increased by experimental conditions of raised temperature, intensified radiation and the application of mutagenous chemicals.

Many amino-acids are coded by more than one nucleotide codon (e.g. leucine is coded by AUU, GUU, and CUU) so a mutation which strikes only one nucleotide might produce no effect at all. Often, however, such a change would upset protein synthesis by introducing a different amino-acid into the molecule. This substitution might severely affect the cell and, if the results were not lethal, this mutation would be handed on to further cell generations. Should a mutation occur in the reproductive cells, then the offspring of that individual would carry the altered characteristics in every cell. Such inherited variations are the raw material from which the evolutionary mechanism builds the complexities of the living world.

5.4   Coding patterns — DNA to DNA

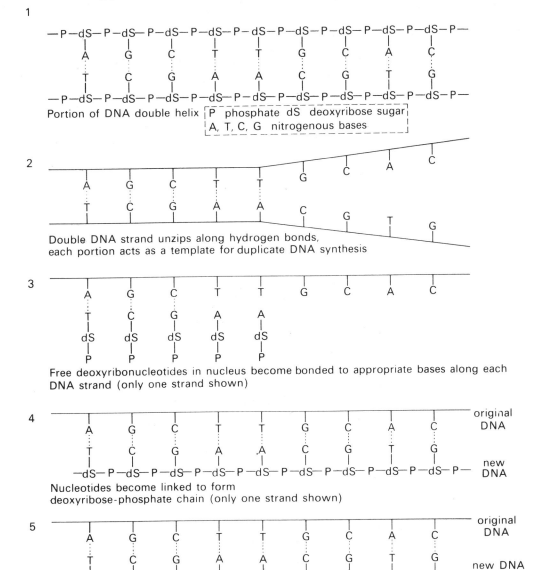

1

—P—dS—P—dS—P—dS—P—dS—P—dS—P—dS—P—dS—P—dS—P—dS—P—

     A   G   C   T   T   G   C   A   C

     T   C   G   A   A   C   G   T   G

—P—dS—P—dS—P—dS—P—dS—P—dS—P—dS—P—dS—P—dS—P—dS—P—

Portion of DNA double helix   P phosphate dS deoxyribose sugar
A, T, C, G nitrogenous bases

2

Double DNA strand unzips along hydrogen bonds,
each portion acts as a template for duplicate DNA synthesis

3

Free deoxyribonucleotides in nucleus become bonded to appropriate bases along each
DNA strand (only one strand shown)

4

Nucleotides become linked to form
deoxyribose-phosphate chain (only one strand shown)

5

Two DNA double chains are completed, identical
to each other and to parent' chain

5.5   *Gene mutations produced by the insertion or deletion of a nucleotide*

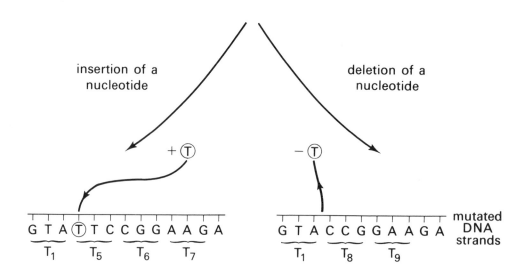

T = a nucleotide codon

Carbohydrate synthesis

Glucose, derived directly or indirectly from photosynthesis in autotrophic cells, is the starting point for the biosynthesis of cellular carbohydrates. Pentose sugars are derived from glucose via the 'pentose shunt'. Hexose sugars are also formed from glucose and most polysaccharides are made from the appropriate pentose or hexose units.

Lipid synthesis

Lipid synthesis requires a good deal of energy so fat stocks are only laid down in times of plenty. Fatty acids are derived from acetyl co-enzyme A and glycerol from the dihydroxyacetone phosphate of glycolysis (see diagram 4.8).

Biosynthesis and the cellular structure

The sites of many biosynthetic activities remain a matter of speculation but some processes have been linked to appropriate structures in the cell.

Since DNA acts as its own synthetic template it must be made in areas of pre-existing DNA which will be mainly in the nucleus. Similarily most types of RNA are manufactured on a DNA template, ribosomal RNA being made specifically in the nucleoli.

Proteins are synthesised from amino-acids on the cytoplasmic ribosomes. Those for intracellular use are made on the ribosomes and polysomes which lie freely in the cytoplasm but proteins due for export from the cell are built on the ribosomes of the rough ER. This allows the proteins to pass along the channels of the ER to the smooth ER and the Golgi apparatus where they may be modified by incorporation into complex lipoproteins or glycoproteins before they are packaged into lysosomes or into secretion vesicles to be discharged from the cell.

Fatty acid synthesis seems to be associated with the cellular membranes, particularily the smooth ER, while some complex polysaccharides and lipids are made in the Golgi apparatus.

FURTHER READING

General Texts
Clowes, R., *The Structure of Life*, 1967, Penguin
Loewy, A. G. and Siekevitz, P., *Cell Structure and Function*, 2nd edition, 1969, Holt, Rinehart and Winston
Watson, J. D., *The Molecular Biology of the Gene*, 1965, Benjamin

Scientific American Offprints
Allfrey, V. G. and Mirsky, A. E., How Cells Make Molecules, Sept. 1961
Crick, F. H. C., The Genetic Code III, Oct. 1966
Fox, C. F., The Structure of the Cell Membrane, Feb. 1972
Kornberg, A., The Synthesis of DNA, Oct. 1972
Nirenberg, M. W., The Genetic Code II, Mar. 1963
Nomura, M., Ribosomes, Oct. 1969
Rich, A., Polyribosomes, Dec. 1963
Yanofsky, C., Gene Structure and Protein Structure, May 1967

6 The Cell at Work: 2 Growth and Division

The cell uses energy to control its internal environment, to maintain its structure and function by biosynthesis (see chapter 5) and for other important cellular processes such as growth and division.

The cell cycle

The period from the formation of a cell to its division into two offspring is known as the cell cycle. For the properties of a cell to remain constant through successive cycles each daughter cell must be endowed with a DNA complement identical to that of its cellular parent. DNA replication and its distribution equally between offspring cells are, therefore, important aspects of division, but the duplication and apportionment of other cellular components and the physical division of the parent cell are integral parts of the process.

The cell cycle has three main stages. First, a period of growth, the interphase, when general cellular synthesis takes place. Then the nuclear division of mitosis, with the separation and redistribution of the chromosomes into two equal daughter nuclei, rapidly followed by the division of the cytoplasm into two daughter cells during cytokinesis.

Cell Growth

The cell grows unevenly during the interphase, which is divided into three periods: G_1, S and G_2 (see diagram 6.1).

G_1 follows cell division and is a period of intensive cellular synthesis. The nucleolus grows rapidly in size and produces ribosomal RNA. Messenger RNA, transfer RNA and ribosomes are manufactured. The synthesis of RNA leads to the production of structural and enzymatic proteins. These enzymes control the metabolic activities involved in the construction of new cellular components for which large quantities of raw materials and energy are needed. G_1 varies greatly in length.

6.1   *The cell cycle*

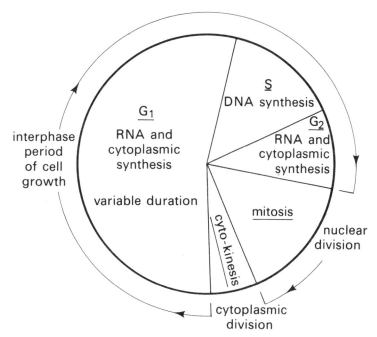

DNA synthesis, which doesn't occur during G_1, takes place throughout the following S period. During this time the cell complement of DNA is doubled and histones (proteins associated with DNA) are also synthesised. Each chromosome duplicates into two chromatids at the end of this stage.

When DNA synthesis is complete the G_2 period of general cellular synthesis continues until mitosis begins. The S and G_2 periods are fairly constant in length for each cell type.

RNA and proteins are synthesised throughout the interphase but the rate is most rapid during the G_1 period of very active cytoplasmic synthesis. Precursors for the mitotic spindle (see page 92) are made during the interphase, mitochondria and chloroplasts also reproduce by division at this time. The centrioles, if present, replicate during the S period by a form of budding.

Mitosis

Cell division usually results in the production of two identical daughter cells from one parent cell. When this occurs the nucleus normally divides by mitosis, a process which distributes the chromosomal material equally between daughter nuclei and ensures that they exactly match the parental nucleus (as compared with meiosis — a special type of nuclear division occurring during gamete formation — which produces offspring cells with only half the parental chromosome content, see page 100).

Mitosis is a complex process because it has to perform a very precise job. The material of each chromosome, duplicated during DNA synthesis in the interphase and divided longitudinally into two chromatids, must be separated to ensure that each daughter cell is furnished with equal chromosome portions. To achieve this the bisected chromosomes must move into an exact arrangement on the mid-plane of a special framework, the spindle, and then the chromatid pairs must separate and move in two bunches to opposite spindle poles so that new nuclear membranes will enclose identical daughter nuclei.

For descriptive convenience, mitosis is always divided into four consecutive phases although the process is quite continuous. The phases are:

A. Prophase — the preparatory phase in which the spindle is formed.
B. Metaphase — during which the chromosomes become attached to the equator of the nuclear spindle.
C. Anaphase — during which duplicate chromatids move to opposite poles of the spindle.
D. Telophase — with the formation of daughter nuclei round each set of chromatids.

The events of mitosis

A. Prophase

1 The chromosomal nucleoprotein, which is dispersed in the interphase cell, becomes condensed into thread-like structures which are visible with the light microscope. These chromosomes are longitudinally divided into two identical chromatids formed during the interphase.

2 The chromosomes condense further, becoming short and stout as each slender chromatid coils like a spring.

3 The nucleoli reduce in size.

4 In animal cells the pair of centrioles (each already replicated during interphase) separate to form the poles of the spindle, an ellipsoid structure made of microtubules. Centrioles have not been seen in higher plant cells where the spindle develops without their help.

5 The nuclear membrane breaks down.

B. Metaphase

1. The chromosomes move to the spindle equator and each one becomes attached to the spindle tubules by a genetically inert area — the centromere.

2. The nucleoli often completely disappear.

C. Anaphase

1 The centromeres now divide so that each chromatid has its own centromere.

2 The component chromatids of each chromosome move to opposite poles of the spindle — the mechanism causing this movement is unknown.

D. Telophase

1 The nuclear membrane reforms round each daughter nucleus.

2 As the nuclear membrane develops the chromatids uncoil, elongate and become indistinct, returning to their active state. The spindle disappears.

3 The nucleoli are reorganised on the nucleolar organisers during this phase.

Cytokinesis

Cytoplasmic division is usually closely coordinated with mitosis. In animal cells, protozoans and many algae, division occurs by 'furrowing'. Here the plasma membrane around the cell circumference moves inwards to form a furrow in the plane of the spindle equator which eventually cuts the cell in two by constriction. A membranous cell plate develops in the plane of the spindle equator of most plant cells instead of the furrow. The plate spreads outwards until it meets the cell wall and it separates the cell into equal portions. A new cell wall will then be laid down on the cell plate.

In both plant and animal cells cytokinesis begins after the chromosomes have moved to the spindle poles during anaphase and it divides the cytoplasmic contents more or less equally between the two daughter cells. It usually takes rather longer in plant than in animal cells and its completion marks both the end of cell division and the end of the cell cycle.

FURTHER READING

General Texts
Ambrose, E. J. and Easty, D. M., *Cell Biology*, 1970, Nelson
Mitchison, J. M., *The Biology of the Cell Cycle*, 1971, Cambridge

Scientific American Offprint
Mazia, D., How Cells Divide, Sept. 1961

7 Cell Types

The generalised cell is a useful notion but most cells are specialised to suit their individual needs.

Procaryotic cells and eucaryotic cells

Cells may be classified structurally into two main groups, the procaryotes and the eucaryotes. Procaryotic cells are those of simple organisms such as the blue-green algae and bacteria. Unlike the generalised cell, the genetic material of a procaryotic cell is not enclosed in a nuclear membrane so the nucleus is often difficult to identify with the usual techniques. While complex infoldings of the plasma membrane may occur, membrane-bounded organelles, such as mitochondria and chloroplasts, are never found and these cells are usually enveloped in a rigid cell wall (see table 7.1).

The eucaryotes include algae, fungi, protozoans and the higher plant and animal cells. They are well represented by the generalised cell, possessing both nuclear membrane and cytoplasmic organelles. This cell type is the basic structural unit of all organisms except the procaryotes and the viruses (viruses are discussed fully in chapter 8).

Plant and animal cells

The eucaryotic cells of higher organisms can be divided into two categories — plant cells and animal cells. The possession of a cellulose cell wall and chloroplasts consigns a cell to the plant group while the absence of both these characters usually places it in the animal division.

In general, a plant cell is functionally an autotrophic cell (see page 60) or forms part of a multicellular organism with specialised photosynthetic structures which allows the whole organism to function autotrophically (some parasitic and saprophytic plants are exceptions to this rule). Animal

Table 7.1 *Some characteristics of procaryotes and eucaryotes*

Feature	Procaryotes	Eucaryotes
Nucleus	No membrane-enclosed nucleus. Genetic material consists of one or more folded DNA strands without associated histones	Enclosed in nuclear membrane. Genetic material present as more than one chromosome. DNA associated with histones (only 40 – 50% DNA).
Cell wall	Cell wall present in all but the smallest procaryotes	Variable
Plasma membrane	Present	Present
ER	Not present	Universally present
Organelles	Membrane-bounded organelles not found	Present
Photosynthetic apparatus	May be present. May contain chlorophyll. Never surrounded by membrane to form organelle	May be present as chloroplast organelle
Ribosomes	Present, free in cytoplasm	Present, free in cytoplasm or associated with ER
Centrioles	Absent	Present in animal cells
Flagella	Sometimes present, made up of single fibre	Sometimes present, characteristic structure of nine sets of micro-tubules arranged in cylinder
Size range	100 – 2,000 nm	10,000 – 100,000 nm

cells are always functional heterotrophs as they never possess the photosynthetic apparatus for harnessing light energy (see table 7.2).

Specialised cells

Many cells in multicellular organisms become highly specialised in order to perform essential functions efficiently. Among these specialised cells are transport cells, assembly cells, sensory cells, nerve cells and a great many more. Three examples are chosen to illustrate cellular specialisation in the multicellular animal; these are the muscle cell, the nerve cell and the gametocyte.

1 The mammalian muscle cell

The muscle cell is an effector cell capable of performing mechanical work by contraction in response to a stimulus.

Structure   Voluntary muscles can produce sudden contractions which bring about relative movements of parts of an organism. They contain striated muscle cells arranged in bundles to make up the muscle fibres. Each elongated multinucleate cell is encased in a strong outer membrane, the sarcolemma. Tightly-packed longitudinal myofibrils, interspersed with many mitochondria, fill the cell. The electron microscope shows that each myofibril has alternating light and dark bands which coincide with those of neighbouring myofibrils to produce the striated appearance of the muscle fibre.

The striated myofibril bands correspond with protein variations along the length of the fibril. Longitudinal filaments in the myofibril may contain either the protein myosin or the protein actin. Where the myofibril is made up of thicker myosin filaments it appears as the darker 'A' band (see diagram 7.1), where it contains only the thinner actin filaments it looks pale, as in the 'I' band. Where these filaments overlap intermediate shades are developed, such as in parts of the 'A' band region (except for the 'H' zone), and cross-links are established between the two filament types.

Function   When a muscle cell contracts, the thin actin filaments slide between the myosin filaments, cross-links breaking and reforming in sequence, until the 'H' zone and the 'I' band disappear (the 'A' band length remains unchanged, see diagram 7.1). In this way the total length of the muscle cell decreases.

Table 7.2   *Some characteristics of plant and animal cells*

Feature	Plant cell	Animal cell
Cell wall	Cellulose cell wall usually present	Cell wall sometimes present, but not made of cellulose
Chloroplasts	Present in photo-synthetic cells	Absent
Vacuoles	Often large and conspicuous in mature cells, may take up 80% of cell volume. Surrounded by tono-plast membrane	If present usually small
Centrioles	Seen in algal cells but not in higher plant cells	Present

7.1 The muscle cell (myofibril)

The relaxed myofibril

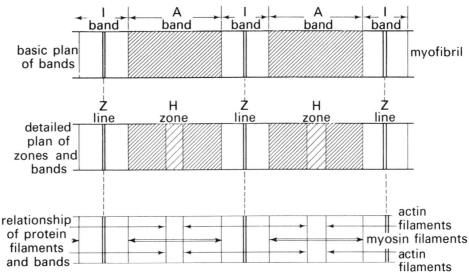

The contracted myofibril

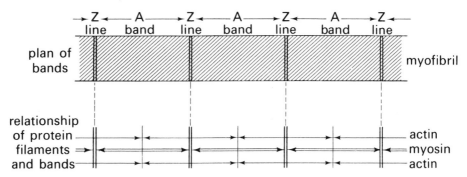

The biochemistry of muscle contraction is not completely understood, but it is known that ATP must be available if contraction is to occur and that the energy of the ATP molecule is used to power the establishment and subsequent breakdown of successive actino-myosin links. An additional energy store is found in vertebrate muscle. It is in creatine phosphate (phosphocreatine) which, like ATP, possesses a high-energy phosphate bond and can be stored in greater quantities in the cell. When the ATP requirement is large, creatine phosphate can donate its phosphate group to ATP.

2 The mammalian nerve cell

Multicellular organisms require intercellular communication systems in order to co-ordinate their varied activities. Two methods of communication exist in higher animals. The slower method is a wholly chemical system in which hormones, produced in discrete regions, travel throughout the organism to exert a specific effect upon another organ. The nervous system is only partially chemical and is a quicker method; rapid local chemical stimulation is produced by means of electrical impulses travelling along inter-connecting pathways. Nerve cells act as links in this system and they are specialised for the unidirectional conduction of electrical impulses.

Structure   The cytoplasm of the nerve cell is drawn out to form one or more nerve fibres while most of the cytoplasmic organelles are contained in a discrete portion of the cell, the cell body, together with the nucleus. A motor nerve cell, transmitting impulses from the brain to the muscles, has its cell body near to the central nervous system and one extremely long process, the axon, extends from the cell body to the muscle it innervates (see diagram 7.2). The axon contains microfilaments, microtubules and mitochondria and may be insulated by a fatty myelin sheath. One nerve cell can make connections with many other fibres though these links are indirect as a small gap, the synapse, always occurs between consecutive fibres.

Function   The nerve impulse is an electrical phenomenon but it is not produced by a flow of electrons. Instead it is propagated by a wave of permeability changes in the usually polarised cell membrane and the conse-quent movement of ions into and out of the nerve fibre.

All cell membranes are selectively permeable and nerve cells, like many others, tend to concentrate potassium ions, K^+, within them and to exclude sodium ions, Na^+. This is achieved by the active 'sodium pump' mechanism which works against the concentration gradient (see page 80). K^+ tend to diffuse back out of the nerve fibre more rapidly than Na^+ can diffuse into it since the membrane is more permeable to K^+ than to Na^+. This leads to a build-up of positive ions on the outside of the fibre and to the development of a negative charge inside the fibre due to the loss of K^+. These effects produce an electrical potential across the membrane called the 'resting potential' (see diagram 7.2).

When an impulse passes along a nerve fibre the permeability of the cell membrane alters so that Na^+ can flood into the cell, first cancelling the electrical potential by depolarisation and then causing it to reverse. The area of membrane just in front of the impulse is stimulated by this change to allow the entry of Na^+ and so these variations in permeability are propagated along the nerve fibre. After the rapid Na^+ inflow there is a sudden outflow of K^+ which restores the original potential difference across the membrane.

7.2 The nerve cell

(a) General structure of a nerve cell with a myelinated axon

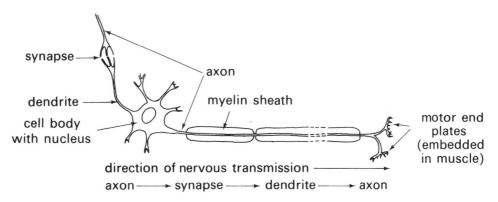

direction of nervous transmission ⟶
axon ⟶ synapse ⟶ dendrite ⟶ axon

(b) Nervous transmission along an axon

direction of transmission of nervous impulse

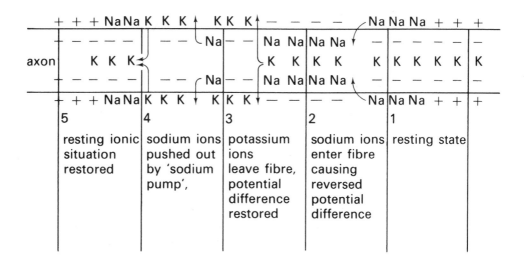

	5	4	3	2	1
	resting ionic situation restored	sodium ions pushed out by 'sodium pump',	potassium ions leave fibre, potential difference restored	sodium ions enter fibre causing reversed potential difference	resting state

The nerve fibre is ready to transmit another impulse as soon as the initial ionic balance is restored. Restoration will require the help of the 'sodium pump' after the passage of many impulses has led to an appreciable loss of K^+. Nervous transmission is very rapid; impulses can travel at 100 metres per second along certain nerves although the rate is far slower along non-myelinated fibres.

A totally different mechanism enables impulses to jump the synaptic gap between fibres or to cross the neuro-muscular junction. Here a chemical transmitter provides the connecting link. The arrival of a nerve impulse at the fibre terminal stimulates the terminal to secrete acetyl choline. This chemical diffuses across the synapse and interacts with a receptor site on the subsequent nerve fibre, producing changes in the membrane permeability to Na^+ and K^+ which induce the propagation of an impulse. Acetyl choline is then rapidly destroyed by the enzyme acetyl choline esterase. Other chemicals such as noradrenalin may also act as synaptic transmitters.

3 The gametocyte of higher animals

Organisms reproducing sexually give rise to new individuals by the fusion of parental gametes. Each species manages to maintain a stable chromosome number in spite of gamete fusion because a special type of cell division called meiosis halves the chromosome number during gamete formation so that fertilisation only restores the parental situation. The gametocyte is a particular cell in the reproductive organs of higher animals which undergoes meiotic division.

Structure   The gametocyte is not obviously structurally specialised and the mechanism which causes it to divide by meiosis rather than by mitosis is unknown.

During gametogenesis the immature sex cells — spermatogonia (male) and oogonia (female) — divide mitotically to produce the gametocytes, spermatocytes and oocytes. These divide meiotically to give rise to the male gametes, the spermatozoa, and the female gametes, the ova, which have a high degree of structural specialisation.

Function   Meiotic division essentially involves one chromosomal duplication followed by two cellular divisions. The sequence of events during meiosis is similar to that of two consecutive mitotic divisions with the important exception that the first prophase is very complex. It involves the pairing of homologous chromosomes on the spindle equator then their separation and movement to opposite poles of the spindle during the first anaphase. The

final products of meiotic division are four haploid daughter cells (see diagram
7.3). In spermatogenesis all the four meiotic products develop into mature
spermatozoa but only one of them matures to form an ovum during oogenesis.

7.3   *The nucleus in division — mitosis and meiosis*

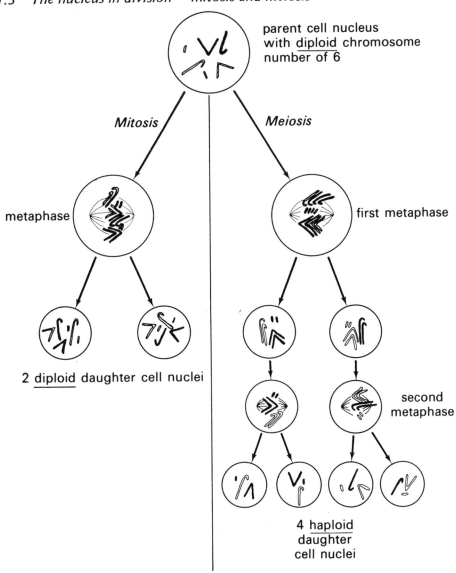

parent cell nucleus
with <u>diploid</u> chromosome
number of 6

Mitosis

Meiosis

metaphase

first metaphase

2 <u>diploid</u> daughter cell nuclei

second
metaphase

4 <u>haploid</u>
daughter
cell nuclei

The events of meiosis

The pre-meiotic interphase is very similar to the mitotic interphase in which cellular synthesis and chromosomal replication occur.

First meiotic division

A. Prophase I   This is a very complex stage which is subdivided for convenience.
   1 Leptotene   The chromosomes become visible. They appear longer and thinner than those of mitotic prophase and, although they are already divided into chromatids, this is not easily seen.
   2 Zygotene   The chromosomes move together into homologous pairs arranged very exactly so that genes are matched along the chromosomes. These pairs are called bivalents.
   3 Pachytene   The bivalents shorten and thicken by coiling.
   4 Diplotene   As attractive forces between chromosomes diminish they move apart. Each chromosome is now seen to consist of two chromatids. Chiasmata (cross-links) are established between the chromatids of homologous partners so that genetic exchange (crossing-over) occurs between homologous chromosomes. Usually even the shortest bivalent develops one chiasma while longer chromosomes form several. Cell growth often occurs with some uncoiling of the chromosomes and some RNA synthesis. This stage may last a very long time and cell growth is most noticeable in oocytes.
   5 Diakinesis   Now chromatids contract still further, the nucleoli disappear, the nuclear membrane breaks down and the spindle forms with centrioles at the poles.

B. Metaphase I   The bivalents become orientated on the spindle equator by their centromeres.

C. Anaphase I   The two homologous members of the bivalent separate and move along the spindle tubules to opposite poles, dragged by their centromeres.
D. Telophase I   Nuclear membranes form round the daughter nuclei, each of which contain half the usual chromosome number (the haploid number). The chromosomes uncoil and cytoplasmic division occurs by furrowing.

E. Interphase   No chromosome replication occurs during this interphase which is of variable length. Sometimes it fails to take place at all and the chromosomes pass directly into the second prophase.

Second meiotic division

F. Prophase II   If an interphase has occurred then the chromosomes reappear and the nuclear membrane breaks down. The spindle forms.

G. Metaphase II   The chromosomes, already divided into chromatids, become attached to the spindle equator by their centromeres.

H. Anaphase II   Centromeres separate and pull the two chromatids of each chromosome to opposite poles of the spindle.

I. Telophase II   The chromatids uncoil, the nuclear membrane reforms, the nucleoli reappear and cytoplasmic division occurs, producing a total of four daughter cells with the haploid chromosome number.

The significance of meiosis

When environmental changes take place a species of organisms must either adapt by evolving or perish. A species can only evolve as a result of the pressures of selection acting upon the inheritable variations that occur among the individual members of a population. Variety is not only the 'spice of life', it is the raw material of evolution.

Inheritable variations in an asexually reproducing species can arise solely by chance mutations but the short life cycles and the rapid reproduction rate of most asexual organisms ensure an adequate supply of variants from this single source. This would not be the case among higher multicellular organisms which have longer life spans and smaller populations. For these species, sexual reproduction and its concomitants of meiotic cell division and fertilisation provide additional sources of inheritable variation.

Mutations produce actual changes in gene structure while sexual reproduction produces different combinations of genes. These may arise in three ways. Homologous chromosomes may swop genes during the cross-over exchanges of genetic parts at meiotic diplotene. Genes may be variously assorted into haploid chromosome groupings when homologous chromosomes move randomly to either daughter nucleus at first meiotic division. Allelic genes may combine in diverse ways during fertilisation when random gametes from each parent chance together to restore the diploid chromosome number. Genetic variation resulting both from these mechanisms and from mutations make it extremely unlikely that any sexually-derived offspring will be genetically identical with either parent. Meiosis therefore provides an important source of variation within a species.

FURTHER READING

General Texts
Ambrose, E. J. and Easty, D. M., *Cell Biology*, 1970, Nelson
Kemp, R., *Cell Division and Heredity*, 1970, Arnold
Novikoff, A. B. and Holtzman, E., *Cells and Organelles*, 1970, Holt, Rinehart & Winston
Swanson, C. P., *The Cell*, 1969, Prentice Hall

Scientific American Offprints
Eccles, J., The Synapse, Jan. 1965
Huxley, H. E., The Contraction of Muscle, Nov. 1958
Katz, B., How Cells Communicate, Sept. 1961

8 Viruses—The Cell Parasites

Viruses are a group of extremely small agents which infect plant and animal cells. They usually show their presence by causing disease and they are unable to multiply outside their host tissue.

Viruses have been aptly described as a bunch of 'genes in search of a cell that will enable them to reproduce'. Certainly their structure is unusual, consisting only of a nucleic acid core wrapped up in a protein coat. They possess none of the recognisable organelles found in a typical cell and so cannot carry out normal cellular functions. In order to compensate for these deficiencies they are highly specific, obligatory, intracellular parasites.

Varieties of viruses

Many hundreds of virus varieties have been identified which parasitise many different cell types and provoke a wide range of diseases. These include infections of agricultural and medical importance such as 'potato leaf roll', 'sugar beet yellows', 'foot and mouth' disease of cattle, 'myxomatosis' of rabbits, and the common 'cold', 'influenza', 'poliomyelitis', 'smallpox', 'measles', 'yellow fever' and 'mumps' in man. Most viruses are specific to a certain cell type within their host, like the poliomyelitis virus which attacks

Table 8.1   *Viruses — a relative size scale*

Structure	Average size range (maximum dimension)
Mammalian cell	5 000 – 50 000 nm
Bacterial cell	500 – 2 000 nm
Mitochondrion	approx. 1 000 nm
Virus	10 – 500 nm
Ribosome	20 – 25 nm

human nervous tissue and the common cold virus which invades the mucous membrane cells.

Classification of viruses

Viruses are submicroscopic so, until recently, their classification was an unrewarding task frustrated by the lack of information about their structure. Many older classifications were based upon the type of disease caused by viruses but such schemes were invalidated by viral properties which have come to light with modern techniques.

The most acceptable of the recent classifications (proposed by the Provisional Committee for the Nomenclature of Viruses) uses structural and biochemical criteria. It divides the virus phylum into two sub-phyla on the basis of the nucleic acid contained in each virus, as viruses may contain DNA or RNA but never both. Within these sub-phyla, divisions are made on grounds of structure (see table 8.2). The three structural types differ in

Table 8.2   *A classification system for viruses*

Phylum	Sub-phylum	Class	Examples of class members
	Deoxyvira (viruses containing DNA)	Helical symmetry	Smallpox and cowpox viruses
		Cubical symmetry	Adenovirus causing croup etc. Herpes virus causing cold sores Chickenpox virus
VIRA		Binal symmetry (combination symmetry)	T-phages which attack bacteria
	Ribovira (viruses containing RNA)	Helical symmetry	Many plant viruses causing mosaic diseases Influenza virus, mumps virus, measles virus and rubella virus
		Cubical symmetry	Polio virus, cold viruses, foot and mouth disease virus

their modes of symmetry. These are exemplified by the helical symmetry of the tobacco mosaic virus (TMV), the cubical symmetry of the regular icosahedron of the adenovirus and the combined symmetries of the complex T-phages (see diagram 8.1).

8.1   Virus types

Symmetry	Example	Structure
cubical	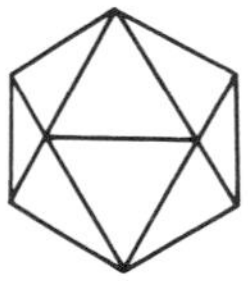 adenovirus 50 nm	An icosahedron with 20 faces made up of 252 protein sub-units. Core made of double-stranded DNA.
helical	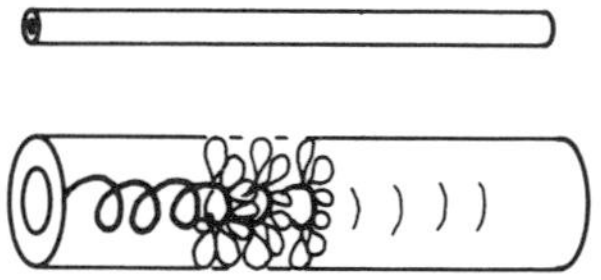 tobacco mosaic virus (TMV) 300 nm long	Tubular structure made up of helical RNA strand with attached protein sub-units.
binal	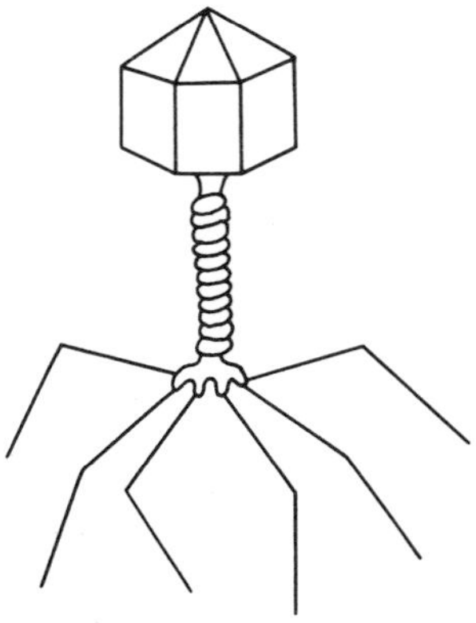 T$_2$ bacteriophage 200 nm	Structure made up of two parts — a headpiece containing DNA core and a contractile tailpiece with six fibres. The whole virus is covered by a protein coat.

108 Cell Biology

Viruses in action

The first step in a viral attack is the attachment of the virus particle (virion) to the host cell membrane. This is followed by the injection of the viral nucleic acid into the cytoplasm of that cell. The intricate protein coat (capsid) of the virus usually remains outside (see diagram 8.2).

8.2 *Viruses in action — a bacteriophage attack*

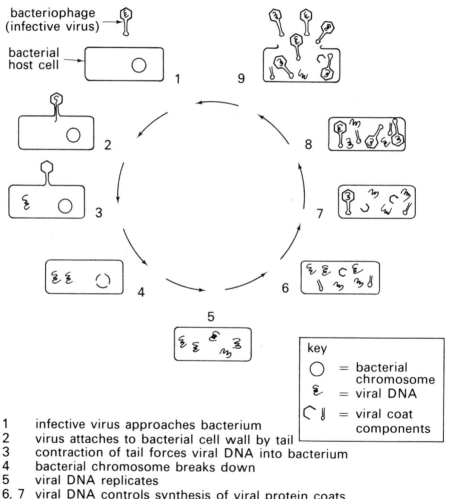

1   infective virus approaches bacterium
2   virus attaches to bacterial cell wall by tail
3   contraction of tail forces viral DNA into bacterium
4   bacterial chromosome breaks down
5   viral DNA replicates
6, 7 viral DNA controls synthesis of viral protein coats
8   new bacteriophages formed
9   host cell ruptures, releasing new bacteriophages to attack
     susceptible bacteria

Once within the host cell, the viral nucleic acid usurps control from the host nucleic acid and proceeds to use host nucleotides and ATP to synthesise more viral nucleic acid. This done, the newly formed viral nucleic acid provides a template for viral protein production from host amino-acid supplies. Capsids for the new viral nucleic acid cores are made from these proteins and eventually the host cell ruptures, releasing fresh virions which will repeat the parasitic life cycle.

This growth cycle is characteristic of the bacteriophages (viruses which attack bacteria) containing DNA. Viruses may contain either DNA or RNA, occurring in double or single strands. The majority of animal viruses possess DNA and have life cycles similar to that of the bacteriophage except for a longer time span and less dramatic host cell breakdown. All known higher plant viruses, some small animal viruses and a few bacteriophages contain RNA instead of DNA. Their life cycles, while differing in some biochemical respects, are essentially similar to that of the DNA-containing bacteriophage.

The temperate phage

While most bacteriophages pursue the reproductive type of life cycle described in the last section, some of them may also exist in a different and more benevolent relationship with their host cells. Under certain conditions the genetic material of one of these 'temperate' phages may become an intimate part of the chromosome of the host bacterium and be transmitted with it to subsequent bacterial generations. The phage genes, nevertheless, retain their identity and may, from time to time, enter into the normal reproductive phase of the viral life cycle. The temperate phage is called a 'prophage' and its host is a 'lysogenic' bacterium. Sometimes the presence of a prophage confers novel properties upon its lysogenic host which are not encountered in the uninfected bacterium.

Some tumour-inducing viruses behave rather similarily since they do not destroy their host cell but alter its functional and developmental patterns; however the true lysogenic situation has been found only among the bacterio-phages.

Viral genetics

Genetics — the study of similarities and differences between parent and offspring — always an interest in agricultural communities, became truly scientific with the work of Gregor Mendel. Farmers pondered over the quality of successive generations of livestock and crop plants, Mendel studied a patch of garden peas and T. H. Morgan observed a rapid progression of fruit fly generations. Nowadays many geneticists concentrate upon even smaller fry — bacteriophages. Indeed the bacteriophage might have been designed to answer a geneticist's plea for enormous numbers of simple organisms with a generation time so short that it can be counted in minutes.

Complex organisms possess many chrosomes (man has forty-six) and consequently large numbers of genes. The role of a single gene in such a cell may be difficult to investigate because of the masking effects of other genes, particularily dominant alleles* in diploid cells. Viruses, together with bacteria, have only one chromosome which makes the task of investigating single gene effects very much easier.

Mutated 'marker' genes with easily discernible effects have been used in virus populations to provide much information about gene action. Viral genetic maps† can now be made as new combinations of viral nucleic acid result from cross-overs between the matching chromosomes of identical viruses simultaneously infecting the same host cell. This process, analogous to chiasmata formation during the first meiotic prophase, allows viral genes to recombine into new genetic groups. It is not known whether recombination acts as an important natural mechanism in the evolution of a viral species but recombination frequency data has allowed detailed genetic maps of viral chromosomes to be developed. Recombination has been demonstrated in some animal viruses but most of this work has been done on bacteriophages which, by reason of their simplicity, convenience and rapid reproduction rate, have proved to be very useful experimental genetic material.

* An allele is one of a group of alternative genes any of which may occupy a certain position on a chromosome. In a diploid organism there are two genes for each character, one on each homologous chromosome. If these two genes are different alleles, usually one allele, the dominant one, will exert its full effect on the character while its partner allele, the recessive one, will not be expressed. Dominant alleles therefore mask the presence of recessive alleles.

† Genes lying on the same chromosome tend to be inherited together as a bunch (linked) except where they are separated by chiasmata during meiosis. Chiasmata are more likely to separate distant genes than near neighbours. Genetic maps which indicate the relative position of genes on a chromosome are usually made by determining the frequency with which two genes on a chromosome become separated by 'crossing-over' during meiosis and recombined into different gene groups.

FURTHER READING

General Texts
Clowes, R., *The Structure of Life*, 1967,Penguin
Goodheart, C. R., *An Introduction to Virology*, 1969, W. B. Sanders
Smith, K. M., *Biology of Viruses*, 1965, Oxford Univ. Press
Stent, G. S., *Molecular Biology of the Bacterial Viruses*, 1963, W. H. Freeman

Scientific American Offprints
Benzer, S., The Fine Structure of the Gene, Jan. 1962
Crick, F. H. C., The Genetic Code, Oct. 1962
Edgar, R. S. and Epstein, R. H., The Genetics of a Bacterial Virus, Feb. 1965
Horne, R. W., The Structure of Viruses, Jan. 1963
Jacob, F. and Wollman, E. L., Viruses and Genes, Jun. 1961
Kellenberger, E. The Genetic Control of the Shape of a Virus, Dec. 1966
Nirenberg, M. W., The Genetic Code III, Mar. 1963
Sinsheimer, R. L., Single-Stranded DNA, July 1962

9 The Cell in Review

Our knowledge of cellular structure and function has been greatly increased by the new techniques of investigation developed over the last thirty years. The electron microscope has revealed the complexities of 'protoplasm' and the cellular organelles, while experimental cytologists and biochemists have elucidated the functions and interrelationships of these structures.

The eucaryotic cell is now known to possess a high degree of internal organisation. It is enveloped by a semi-permeable plasma membrane which controls the passage of substances between the cell and its environment. The cell's interior is extensively subdivided into smaller communicating areas by a flexible arrangement of membranes, the endoplasmic reticulum. Membrane-wrapped organelles of different types house the materials and enzymes concerned in the various cellular activities.

The nucleus is important among the cellular organelles as it contains a library of genetic information stored in DNA nucleotide code and a copying mechanism which allows relevant portions of this information to be transcribed into messenger RNA. Protein synthesis in the cytoplasm is controlled from the nucleus by means of messenger RNA blueprints. Ribosomal RNA is also manufactured within the nucleus in dense nucleolar areas.

The photosynthesising chloroplasts of autotrophic eucaryotic cells are of great significance as the whole system of energy utilisation in the living world is based upon the process of photosynthesis. The presence of chlorophyll pigments and enzymes within the elaborate membranes of chloroplasts enables light energy to be used to power carbohydrate synthesis from inorganic supplies of carbon dioxide and water. All cells require organic respiratory fuels, such as glucose, to provide the energy for cellular activities; cells without chloroplasts must obtain preformed respiratory substrates from autotrophic sources outside the cell.

Oxidative respiration, the process which progressively releases the stored energy of complex organic molecules, occurs largely within the mitochondria organelles. This multi-stage cyclical procedure breaks down carbohydrates and

9.1   *Patterns of cell activity*

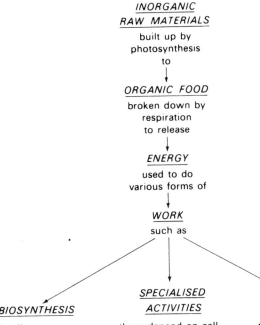

*INORGANIC
RAW MATERIALS*
built up by
photosynthesis
to

ORGANIC FOOD
broken down by
respiration
to release

ENERGY
used to do
various forms of

WORK
such as

BIOSYNTHESIS

all cells must
do this to
some extent:

1 manufacture of
nucleic acids

2 manufacture of
proteins

3 manufacture of
other organic
materials essential
to structure and
function of the cell

*SPECIALISED
ACTIVITIES*

these depend on cell
type and may include:

1 transport of
materials

2 change in shape
such as muscle cell
contraction or
production of pseudo-
podia

3 transmission of
nervous impulses

4 emission of
electrical impulses

5 emission of light

6 formation of special
substances such as
bone, wax etc.

7 other specialised
activities

REPRODUCTION

most nucleate cells
can reproduce by
division:

1 by mitosis,
simple cell
division
or, exceptionally,
2 by meiosis,
specialised cell
division undergone
by some cells

other fuels to carbon dioxide and water, repackaging the energy of the original molecular configuration into the convenient high-energy bonds of the ATP molecule.

The Golgi apparatus, another membranous organelle, is concerned with the secretion and packing of proteins and polysaccharides, and probably also with the synthesis of certain complex carbohydrates.

Lysosomes contain hydrolytic enzymes capable of breaking down organic macromolecules. In order to preserve the cell from wanton self-destruction these enzymes are stored within a protective membrane but their controlled liberation plays an important part in the constant cellular demolition and replacement of worn components and in the digestion of material entering the cell by pinocytosis and phagocytosis.

Specific types of proteins are essential both for cellular growth and repair and for the production of the enzymes which catalyse biochemical reactions. Protein manufacture is controlled by genetic instructions encoded in the nucleus and takes place on the surface of very small granular ribosomes in the cytoplasm. When a certain protein is required by the cell a portion of genetic DNA containing the relevant information is transcribed on to messenger RNA which leaves the nucleus and moves into the cytoplasm. There, lying over a group of co-operating ribosomes, the messenger RNA is held in place by ribosomal RNA and provides a template for specific protein synthesis. Amino-acids are individually transported to their appropriate places on the mRNA template by transfer RNA. This elaborate mechanism ensures that very complex proteins are constructed with a high degree of accuracy and may efficiently perform their particular functions. Proteins for intracellular use are made on the free cytoplasmic ribosomes whereas proteins for export are synthesised on the ribosomes of the rough ER and transported through the ER to the Golgi apparatus where they are packed and dispatched.

Electron micrographs show the structural complexities of the cell in marvellous detail; it is less easy to demonstrate the constant activity that occurs at all levels of cellular function. The cell is a very dynamic unit. Its structure is always undergoing repair and reconstruction. Membranes are constantly being assembled, dismantled and reassembled elsewhere, often at great speed. Membranes are very flexible too. Mitochondria continually change shape and divide, in liver cells they seem able to replicate in response to heavy cellular energy demands. Depletion of light will cause photosynthetic cells to lose both the integrity of their chloroplasts and their supply of chlorophyll but these will quickly be reorganised and reformed when light is restored. This state of constant activity allows the cell to respond swiftly to changing environmental demands — an invaluable attribute.

The tasks ahead

It is tempting to assume that the torrent of research directed at Hooke's cell during the last thirty years must, by now, have revealed every detail of its construction and every biochemical reaction involved in its activities. Such an assumption would be quite wrong. The picture is very far from complete and many unanswered questions intrigue and stimulate investigators today. How, for example, does DNA replication actually take place? What causes the double helix to separate into two strands? Why do the DNA strands sometimes act as templates for the production of more DNA and at other times serve as templates for mRNA synthesis — in which case only one strand of each pair will be active? Exactly how are cellular activities controlled, regulated and co-ordinated? What switching mechanism turns on the genetic commands when their end product is needed by the cell and cuts them off when the product is no longer required?

One of the current theories of cellular control has been developed by Jacob and Monod at the Pasteur Institute in Paris and it will usefully serve as an example of a promising research field. A key factor in cellular control is the regulation of enzyme production and Jacob and Monod sought to explain the lactose-metabolising activities of the *Escherichia coli* bacteria which live in the human gut. These bacteria will produce lactose-splitting enzymes when they grow in the presence of lactose sugar, a property which allows the bacteria to use lactose as a source of energy. If the growth medium contains no lactose, however, the bacteria do not produce the lactose-splitting enzymes. To account for this situation where the presence of the substrate seemed to initiate enzyme production Jacob and Monod developed the 'operon' theory.

Jacob and Monod postulate the existence of three distinct genetic elements which act together to account for enzyme production and regulation. They suggest that the genetic instructions for the synthesis of a group of enzymes acting sequentially in a biochemical pathway will occur in sequence on one DNA strand (there is experimental evidence to support this idea). This 'structural' gene sequence, the first genetic element, would be controlled by an 'operator' gene located at one end of the linked gene chain. The operator gene, the second genetic element, would act as an 'on' switch for its series of structural genes, the whole group of genes being known as an 'operon'. The third genetic element involved in enzyme production is a 'regulator' gene. This may be located at a different site from its operon and it would control the synthesis of a 'repressor' molecule (see diagram 9.2). The repressor molecule is normally in constant production, its presence inhibiting the

116   Cell Biology

9.2   *The Jacob and Monod 'operon' theory as illustrated by lactose metabolism in E coli*

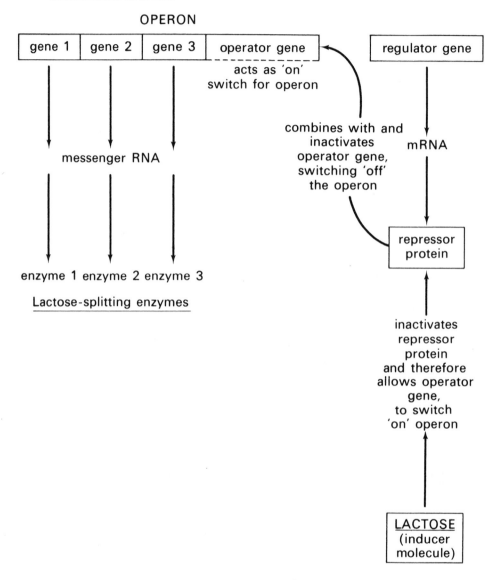

action of the operator gene and so switching *off* the operon. Repressor molecules would be inactivated by specific 'inducer' molecules which would stop repressor action and so, by a double negative process, activate the operon. Jacob and Monod theorise that, in the case of *E. coli*, the lactose molecule itself serves as an inducer molecule by inactivating the repressor molecule for the lactose-splitting enzyme operon. In this way the presence of lactose would initiate enzyme production.

Genetic feed-back mechanisms like Jacob and Monod's operon model could be used to explain many examples of genetic regulation. Such theories stimulate research to determine the extent of their applicability. In other areas yet more questions abound — what causes cells to divide, why do they differentiate, what determines the abnormal behaviour of cancer cells? New techniques develop to test new theories, new information accumulates and cell biology advances step by step.

BOOKS ON CELL BIOLOGY

Low Price Range   less than 60p

Rose,   *The Chemistry of Life* ⎫ Penguin
Clowes, *The Structure of Life* ⎭

Scientific American offprints — list from W. H. Freeman & Co Ltd, 58 Kings Road, Reading RG1 3AA

Oxford Biology Readers — list from Oxford University Press, Press Road, Neasden, London NW10 ODD

Medium Price Range   60p to £2.00
Baron,   *Organization in Plants*, Arnold
Goldsby, *Cells and Energy*, Macmillan
Hurry,   *Microstructure of Cells*, Murray
McElroy, *Cell Physiology and Biochemistry* ⎫ Prentice Hall
Swanson, *The Cell* ⎭

Institute of Biology, *Studies in Biology* — a series, Arnold

Higher Price Range   over £2.00
Ambrose and Easty,   *Cell Biology*, Nelson
Fawcett,   *Atlas of Fine Structure*, Saunders
Loewy & Siekevitz,   *Cell Structure and Function* ⎫ Holt, Rinehart & Winston
Novikoff & Holtzman, *Cells and Organelles* ⎭
Scientific American,   *The Living Cell*, Freeman
Watson,   *Molecular Biology of the Gene*, Benjamin

These books are mentioned in greater detail in the relevant chapter references. All book prices are subject to change.

Index

Figures in italic refer to diagrams

acceptor molecules, *see* carrier
    molecules
acetic acid, 46, *3.9*
acetyl choline, 100
acetyl choline esterase, 100
actin, 96, *7.1*
active transport, 78, 79
adenine, 49, 52, *3.11*, *3.13*, 83
adenosine triphosphate, ATP,
    19, 32, 52, 60, 64, *4.1*,
    69, 114
    synthesis, 68, 75
aerobic respiration, 19, 68, *4.7*,
    4.8, *4.9*, 69, 71, 72–4,
    112
alanine, *3.6*
aldose, 30, table 3.2
algae, 26, 80, 94
    blue-green, 94
allele, 103, 110
allosteric effect, 56
amino-acid, 29, 40, *3.6*, 77, 82,
    89, 114
ammonia, 77
anaerobic respiration, 68, *4.7*,
    69, 71, 73–4
anaphase, 92, 93, 102, 103
araban, 37, table 3.2
arabinose, 32, table 3.2
aspartic acid, *3.6*
autotroph, 60, *4.1*, 88, 112
axon, 98, *7.2*

bacteria, 23, 94, 109, 110
    lysogenic, 109
bile salts, 47
biosynthesis, 82–9
    and cell structure, 89
bivalent, 102
butyric acid, *3.9*

Calvin cycle, *see* carbon
    dioxide fixation

capsid, 108–9
carbohydrate, 28, 29, table
    3.2, 30–9
    classification, 30
    conversion, 63, *4.3*, *4.6*, 66
    derivatives, 37, 39
    as fuel, 75, *4.11*
    phosphorylated, 39, 52
    respiratory breakdown of,
        68–74
    synthesis, 88, 112
carbon dioxide, 69, 72, 75
carbon dioxide fixation,
    Calvin cycle, Dark
    reaction, 63, *4.3*, 66,
    4.5
carbon monoxide, 56
carrier molecules, 52
    acetyl carrier, 72
    carbon dioxide acceptor,
        66, 75
    electron carrier, 64
    energy carrier, 72
    hydrogen carrier, 64, 66, 72
cartilage, 37
cell
    body, 98
    chemistry, 28–59
    cycle, 90–3
    division, 91–3
    generalised, 13–27, *2.2*
    growth, 90–1
    higher animal, 94–5, table
        7.2
    higher plant, 94–5, table
        7.2
    mammalian muscle, 96–7,
        7.1
    mammalian nerve, 98–100,
        7.2
    membranes, 78, 79, 114
    plate, 93
    specialised, 95–104

striated muscle, 96–7
structure, 13–27
structure and function,
    112–14
theory, 11
types, 94–104
wall, 15–16, *2.3*, 93, 94
cellulose, 37, *3.4*, 15, 16, 28
cement, intercellular, 15, 37
centriole, 24, 25, *2.7*, 91, 92,
    102
centromere, 93, 102
cephalin, 47
chiasmata, 102, 110
chitin, 16, 28, 37
chlorophyll, 22, 63–8, 112,
    114
    a and *b*, 64
chloroplast, 22, *2.6*, 48, 64,
    91, 114
    and photosynthesis, 66, 94,
        112
cholesterol, 16, 47
chromatid, 91–3, 102–3
chromosome, 26, 82, 90–3,
    100–103
    mutation, 86
cilia, 24, 25
citric acid, 72, *4.9*
    cycle, *see* Krebs' cycle
codon, 83, 86
    anti-codon site, 84
co-enzyme, 52, 59
    A, 52, 72, 76
    acetyl co-8, A, 72, 76, 77,
        88
    nucleotide, 52
collagen, 42
cotton, 37
creatine phosphate, 97
cristae, 19, 75
crossing-over, 102
cycle, cell, 90–3

cysteine, *3.6*, *3.7*, 42
cytochrome, 56, 72, *4.10*
cytokinesis, 90, 93, *6.1*
cytoplasm, 14, 18, 48
　　division, *see* cytokinesis
cytosine, 49, *3.11*, 52, *3.13*, 83

dark reaction, *see* carbon
　　dioxide fixation
decarboxylation, 72
dendrite, *7.2*
deoxyribose, 32, 49, *3.11*, 52,
　　75
desmosome, 17
diffusion, 78–9
dihydroxyacetone, 32, *3.1*
　　phosphate, 76, *4.8*, 88
dipeptide, 40, *3.7*
diploid number, 26, 110
diplotene, 102
disaccharide, 28, table 3.2, 30,
　　3.3, 32, 35
disulphide bond, 42, *3.7*
DNA, 20, 23, 26, 29, 32, 47,
　　49, 52, *3.12*, 86, 114
　　code, genetic code, 82–4,
　　　　112
　　synthesis, 86, 89, 91

elastin, 42
electron transfer, carrier, 64,
　　72
endoplasmic reticulum, ER,
　　18, 23, 112
　　rough, granular, 18, 89, 114
　　smooth, 18, 89
endosymbiont, 20
energy,
　　conversion, 62
　　currency in cell, 60
　　sources, 75
enzyme, 17, 20, 23, 29, 45, 52,
　　55–9
　　action, 55, *3.14*
　　activators, 59
　　active site, 55
　　activity factors, 56, *3.15*
　　e-controlled permeability,
　　　　79
　　e-substrate complex, 55,
　　　　3.14
　　inhibition, 55, *3.14*
　　regulation of e. production,
　　　　115
Escherichia coli, 115
ester, 45

ethyl alcohol, 69, 71
eucaryote, eucaryotic cell, 18,
　　94, table 7.1, 112

fat, *see* lipid
fatty acid, 17, 29, 46
　　breakdown, 76
　　saturated, 46
　　synthesis, 88, 89
　　unsaturated, 46
ferrodoxin, 64
flagella, 24, 25
flavin adenine dinucleotide,
　　FAD, 72, 75
flavin mononucleotide, FMN,
　　72
flavoprotein, 72, 73, *4.10*
Flemming, Walter, 12
fructose, 32, *3.2*, 35, 37, 39,
　　66
　　6-phosphate, 39, *3.5*
　　1,6-diphosphate, 39, *3.5*
fuels, for cellular respiration,
　　75
fungi, 14, 25, 94
furrow, 93

G_1 period, 90, *6.1*
G_2 period, 90, 91, *6.1*
galactose, 35, *3.2*
gamete, 26, 100
gametocyte, 100
gene, 26
　　code, *see* DNA code
　　map, 110
　　'marker', 110
　　mutation, 86, 103
　　operator gene, 115
　　regulator gene, 115
　　structural gene sequence,
　　　　115
genetic code, *see* DNA code
glucose, 28, 32, *3.2*, 35, 37, 66,
　　68, 76, 88
　　6-phosphate, 39, *3.5*, 69,
　　　　75
glutamic acid, *3.6*
glyceraldehyde, 32, *3.1*
　　3-phosphate, 76, *4.8*
glycerol, 47, *3.9*, 76, 88
glycine, 40, *3.6*
glycogen, 28, 37, *3.4*, 75
glycolipids, 16, 47
glycolysis, 69, *4.7*, *4.8*, 75
glycoside link, 28, 35, *3.3*

Golgi apparatus, complex,
　　body, 18, 23, 89, 114
grana, 22–3, *2.6*, 64, 68
granules, 25
　　basal, 24
guanine, *3.11*, 49, *3.13*, 83

haemoglobin, 45, 56
haploid number, 26, 101
helix, double, *3.12*, 52
heterosaccharide, 37
heterotroph, 60, *4.1*
hexosan, 37
　　storage, 37
　　structural, 37
hexose, 30, table 3.2, *3.2*, 32,
　　35, 88
high-energy bond, 52, 60
histone, 26, 45, 91
homologous pair, 26, 100, 102,
　　103
Hooke, Robert, 11, 115
hormone, 45, 98
　　sex, 47
hydrogen bond, 42, *3.7*, 52

interphase, 90–1, *6.1*, 102
inulin, 32, 37, 66
iodine, 37, 80
ionic bond, 42, *3.7*
isomerism, 32
　　optical, 32, 40
　　structural, 32

Jacob, Francois, 115–17

keratin, 16, 42
ketoacid, 77
keto-glutaric acid, 72, *4.9*, 77
ketose, 31, table 3.2
kinetosome, 24
Krebs' cycle, tricarboxylic acid
　　cycle, citric acid cycle,
　　69, 72, *4.9*, 74, 75, 76

lactic acid, 69, 71, *4.7*, *4.8*
lactose, 35, *3.3*, 115–17, *9.2*
lamella, middle, 15, 16
lamellae, 23
lecithin, 47, *3.10*
Leeuwenhoeck, Anton van, 11
leptotene, 102
leucine, *3.6*
light reaction, *see* photolysis
lignin, 16, 28, 37

lipid, fat, 28, 29, 45–7, *3.9*, *3.10*
  as fuel, 76
  neutral, 47
  structure, 45
  synthesis, 88, 89
lipoprotein, 16, 18, 45, 75, 89
lysosome, 23, 80, 89, 114

macromolecule, 28, table 3.1
maltose, 35, *3.3*
mannose, 35, *3.2*
matrix, intercellular, 16
meiosis, 92, 100–103, *7.3*
  significance of, 103
membrane,
  nuclear, 18, 24, 26, 82, 84, 92, 93, 94, 102, 103
  plasma, 15, 16–17, 18, 94, 112
  'unit', 17, 18, 26
Mendel, Gregor, 110
metaphase, 92, 93, 102, 103
microfilaments, 25, 98
microscope,
  electron, 12, 13, 112
  interference, 12
  light, 11, 12
  phase contrast, 12
microtubules, 25, 92, 98
microvillus, 17
mitochondria, 19–20, *2.5*, 23, 47, 91, 96, 98,
  and cellular respiration, 75, 112, 114
mitosis, 90, 91–3, *6.1*, 100, *7.3*
Monod, Jacques, 115–17
monosaccharide, 28, 30–5, table 3.2, *3.1*, *3.2*, 75
Morgan, T. H., 110
mucilage, 35
mucus, 37
muscle cell, 25, 96–7, *7.1*
mutation, 86, 103
  rate, 86
myelin, 17, 98, 100, *7.2*
myofibril, 96
myosin, 42, 96

nerve,
  cell, 98, *7.2*
  fibre, 17, 25, 98–100, *7.2*
  impulse, 98–100
  resting potential of, 98

nicotinamide adenine dinucleo-tide, NAD, 52, 69, 71, 72, 75
  phosphate, NADP, 52, 64, 68
  reduced, NADH$_2$, 64, 66, 69, 71, 72, 73, 74, *4.10*
nitrogenous base, 29, 49, *3.11*, 83
  pairs, *3.13*, 52, 86
noradrenalin, 100
nuclear envelope, *see* membrane, nuclear
nuclear membrane, *see* membrane, nuclear
nuclear spindle, *see* spindle
nucleic acid, 28, 29, 48–54, 75
  structure, 49, *3.11*, *3.12*, *3.13*
  synthesis, 86, *5.4*
nucleolar organiser, 26, 93
nucleolus, 23, 26, 89, 92, 93, 102, 103, 112
nucleotide, 49, *3.12*, 75
  phosphorylated, 52
nucleus, 25–6, 47, 48, 89, 112, 114

oil, *see* lipid
operon, 115
  inducer molecule, 116, 117
  repressor molecule, 115
  theory, 115
organelles, 18–25, 94, 112
osmosis, 79
  osmotic pressure, 79
ovum, 100, 101
  oocyte, 100
  oogonia, 100
oxaloacetic acid, 72, *4.9*, 77
oxidation/reduction reactions, 64

pachytene, 102
pectin, 15, 28, 37
pellicles, 16
pentosan, 37
pentose, 29, 30, 32, table 3.2, *3.1*, 49, 88
  shunt, 75, 88
pepsin, 58
peptide bond, link, 40, *3.7*
pH, 56

phage,
  in action, 108
  prophage, 109
  T-phage, 107, *8.1*, *8.2*, table 8.2
  temperate, 109
phagocytosis, 23, 80, 114
phagosomes, 80
phosphate, 29, 49, *3.11*
phosphocreatine, *see* creatine phosphate
phosphoglyceric acid, PGA, 66, *4.5*
  phosphoglyceraldehyde, PGAL, 66, *4.5*, *4.6*
phospholipids, 16, 23, 48
phosphorylation, 52, 69
photolysis, light reaction, 63, *4.3*, 64, *4.4*, 68
photosynthesis, 22–3, 28, 32, 62, 63–8, *4.3*, *4.4*, *4.5*, *4.6*
  and chloroplasts, 66, 112
pinocytosis, 23, 80, 114
  pinocytic vesicle, 80
polypeptide, 40
  bonds, 3.7
polyribosome, polysome, 84
polysaccharide, 18, 28, table 3.2, 30, 32, 35–7, *3.4*, 75, 88, 114
  and cell structure, 89
  mucopolysaccharide, 37
potassium ions, 80, 98–100
procaryote, procaryotic cell, 20, 94, table 7.1
prophase, 92, 102–3
protamine, 45
protein, 28, 29, 40–5, 114
  chromoprotein, 45
  classification, 42
  conjugated, 40, 45, 55
  denaturation, 45, 56
  enzymatic, 45, 90
  fibrous, 40, 42
  flavoprotein, 45
  as fuel, 77
  globular, 40, 42, 55
  glycoprotein, 45
  metalloprotein, 45
  mucoprotein, 45
  nucleoprotein, 45
  phosphoprotein, 45
  primary structure, 42, *3.8*
  secondary structure, 42, *3.8*

simple, 42, *3.8*
structural, 42, 90
structure, 40–2
synthesis, 82–5, 89
protozoa, 26, 94
pseudopodia, 80
purine, 49, 52, *3.11*
pyrimidine, 49, 52, *3.11*
pyruvic acid, 69, 71, 72, 74, 77, *4.7, 4.8, 4.9, 4.11*

quinone, 72, *4.10*

recombination, genetic, 110
residual body, 80
respiration,
    aerobic, 19, 68, *4.7*, 69, 70, 72, 112
    anaerobic, 68, *4.7*, 69, 70, 72
    cellular, 20, 62, 68–77
    energy yield, 73–4
    fuels for cellular respiration, 75–7, *4.11*, 114
    and mitochondria, 75
    pathways, 32, 39, 75
respiratory chain, 72, 73, *4.10*
ribose, 32, 49, *3.11*, 75
ribosomes, 18, 20, 23, 48, 82, 84, 89, 90, 114
ribulose diphosphate, RDP, 66, *4.5*, 75
ribulose-5-phosphate, 75
RNA, 23, 26, 29, 32, 47–9
    synthesis, 86
    messenger, 82–4, table 5.1, 90, 112, 114
    ribosomal, 84, table 5.1, 89, 90, 112, 114

transfer, 84, table 5.1, 90, 114

S period, 90–1
sarcolemma, 96
Schleiden, 11
Schwann, 11
    cell, 17
serine, *3.6*
size scale, *2.1*, table 8.1
sodium,
    ions, 80, 98–100
    pump, 80, 100
sperm tails, 25
spermatocyte, 100
spermatogonia, 100
spermatozoon, 100, 101
spindle, nuclear, 24, 91, 92, 93, 102, 103
starch, 28, 37, *3.4*, 66
stearic acid, 46
steroid, 48, *3.10*
stroma, 23, 68
succinic acid, 72, *4.9*
sucrose, 35, *3.3*, 66
sulphanilamide, 56
synapse, 100, *7.2*
    synaptic messenger, 100
synthesis, and cellular structure, 89
    carbohydrate synthesis, 88
    lipid synthesis, 88
    nucleic acid synthesis, 86
    protein synthesis, 82

tautomerism, 86
telophase, 92, 93, 102, 103
tendon, 37
testosterone, *3.10*

tetraploid, 26
thymine, 49–52, *3.11, 3.13*, 83
tonoplast, 25
tricarboxylic acid cycle, *see* Krebs' cycle
triose, 30, table 3.2, *3.1*, 32
triploid, 26
turgidity, 25

units of size, table 1.1
uracil, 49, *3.11*
urea, 77
uric acid, 77

vacuole, 25
    digestive, 80
valine, *3.6*
vesicle, 18
    pinocytic, 80
    secretion, 89
virion, 108, 109
virus, 23, 105–11
    adenovirus, 107, *8.1*, table 8.2
    classification, 106, table 8.2
    genetics, 109–10
    in action, 108–9
    tobacco mosaic, 107, *8.1*, table 8.2
    varieties, 105, *8.1*
Virchow, Rudolf, 11, 12

xylan, 37, table 3.2
xylose, 32, table 3.2

zygotene, 102